Maintenance and Safety of Aging Infrastructure

Maintenance and Safety of Aging Infrastructure

Editor

Vineet Kasana

scitus
academics

Maintenance and Safety of Aging Infrastructure

Edited by **Vineet Kasana**

Printed in 2017

ISBN: 978-1-68117-002-2

Library of Congress Control Number: 2015931814

SCITUS Academics LLC,
616, Corporate Way, Suite 2, 4766,
Valley Cottage, NY 10989

www.scitusacademics.com

Contents

Preface

This book presents the latest scientific research and application practice findings in the engineering field of "maintenance and safety of aging infrastructure." The selected invited contributions will provide an overview of the use of advanced computational and/or experimental techniques in damage and vulnerability assessment as well as maintenance and retrofitting of aging structures and infrastructures for minimization of losses and life-cycle-cost. Cost-competent maintenance and management of civil infrastructure requires balanced consideration of both the structure performance and the total cost accrued over the entire life-cycle. Another major problem is that the structure performance is usually reduced during its functioning due to environmental and other factors. Thus, current structural condition state is usually assessed by visual inspection or more advanced automatic structural health monitoring techniques. Furthermore, maintenance managers often require a list of prioritized maintenance interventions for civil infrastructure on an annual and/or long-term basis. Various unavoidable uncertainties associated with both randomness and imperfect knowledge also play a crucial role in management and maintenance of engineering systems. Taking into account the aforementioned issues, this volume aims to present the recent developments of life-cycle maintenance and management planning for deteriorating civil infrastructure considering simultaneously multiple and often competing criteria in terms of condition, safety and life-cycle cost.

In this context, major topics treated in this book include aging structures, climate adaptation, climate change, corrosion, cost, damage assessment, decision making, extreme events, fatigue life, hazards, hazard mitigation, inspection, life-cycle performance, maintenance, management, NDT methods, optimization, redundancy, reliability, repair, retrofit, risk, robustness, resilience, safety, stochastic control, structural health monitoring, sustainability, uncertainties and vulnerability.

This up-to-date overview of the field of maintenance and safety of aging infrastructure makes this book a must-have reference work for those involved with structures and infrastructures, including students, researchers and practitioners.

Editor

Chapter 1

Long-Term Flexural Behaviors of GFRP Reinforced Concrete Beams Exposed to Accelerated Aging Exposure Conditions

Yeonho Park[1], Young Hoon Kim[2], and Swoo-Heon Lee[1]

[1]Department of Civil Engineering, the University of Texas at Arlington, Arlington, TX 76019, USA.

[2]Department of Civil and Environmental Engineering, University of Louisville, Louisville, KY 40292, USA

ABSTRACT

This study investigates the impact of accelerated aging conditions on the long-term flexural behavior and ductility of reinforced concrete (RC) members with glass fiber-reinforced polymer (GFRP) bars (RC-GFRP specimen) and steel bars (RC-steel specimen). A total of thirty six specimens were designed with different amounts of reinforcement with three types of reinforcing bars (*i.e.*, helically wrapped GFRP, sand-coated surface GFRP and steel). Eighteen specimens were subjected to sustained loads and accelerated aging conditions (*i.e.*, 47 °C and 80% relative humidity) in a chamber. The flexural behavior of specimens under 300-day exposure was compared to that of the companion specimens without experiencing accelerated aging conditions. Results indicate that the accelerated aging conditions reduced flexural capacity in not only RC-steel, but also RC-GFRP specimens, with different rates of reduction. Different types of GFRP reinforcement exhibited different rates of degradation of the flexural capacity when embedded in concrete under the same exposure conditions. Several existing models were compared with experimental results for predicting the deflection and deformability index for specimens. Bischoff and Gross's model exhibited an excellent prediction of the time-dependent deflections. Except for the deformability index proposed by Jaeger, there was no general trend related to the aging duration. This study recommends the need for further investigation on the prediction of the deformability index.

INTRODUCTION

The corrosion of steel reinforcement in concrete is a major deterioration mechanism, leading to degradation in the strength and stiffness of reinforced concrete structures. The reduction in the useful service-life of steel-reinforced concrete structures is a cause of concern to the construction industry, mainly due to reinforcement corrosion [1]. However, to reduce costs for repairing significantly corroded structures, in recent years, non-metallic materials of fiber-reinforced polymer (FRP) are increasingly being used as an alternative to steel reinforcement. The growing application of FRP reinforcement is a result of its resistance to corrosion and high strength-to-weight ratio when compared to conventional steel reinforcement [2].

Significant research has been conducted in the past decade to assess the suitability of FRP reinforcement in reinforced concrete beams. Additionally, FRP application is more common in repair or rehabilitation fields, because it is easily attached to the surface of structures [3,4]. There are typically three popular types of fibers, including carbon, glass and aramid fiber. Recently, hybrid carbon-glass FRP sheets have also been used [5]. Among them, this study focuses on glass-FRP (GFRP) reinforcement as a construction material. Particularly, GFRP reinforcement is widely used for flexural reinforcement. The application is not limited to bridge decks, pavements, walls and other systems exposed to harsh environments. GFRP gained momentum when the American Association of State Highway and Transportation Officials (AASTHO) approved the use of GFRP for bridge deck elements [6]. This approval resulted from significant efforts by engineers and researchers to standardize test methods and designs for structural elements using composite reinforcement [7]. Even though GFRP bars exhibit a high strength-to-weight ratio, the modulus of elasticity of GFRP rods is lower than that of steel. Therefore, the design of concrete reinforced with glass fiber (RC-GFRP) beams is often governed by the serviceability limit state (e.g., the maximum allowable deflection and crack widths) rather than the strength limit state (*i.e.*, flexural capacity) [8,9]. Therefore, the prediction of the deflection in a cracked section is considered the important design and analytical procedure. However, various prediction models were formulated based on conventional steel-reinforced concrete. The applicability of existing models to evaluate flexural behavior is still not fully understood. For example, Branson [8] proposed the concept of effective moment of inertia for concrete beams reinforced with steel to predict the deflection of a cracked GFRP beam. Later, Gao *et al.* [10] proposed a modified expression for the effective moment of inertia in order to account for reduced tension stiffening (*i.e.*, the contribution of concrete for tensile stress after cracking). The modification was adopted in the current American Concrete Institute (ACI) 440.1R-06, "Guide for the Design and Construction of Structural Concrete Reinforced with FRP Bars" [7]. However, the information on the applicability of these models for degraded GFRP beam is essentially unknown. Moreover, regarding ductility, due to the linear elastic properties of the GFRP bars up to failure, the conventional definition of ductility cannot be applied to concrete beams reinforced with GFRP reinforcement. Several research papers on GFRP in concrete structures identified the major measure of pseudo-ductility as the deformability index [11,12,13]. Several evaluation methods, such as the energy-based method and the deformation-based method, have been suggested to calculate the ductility index (also called the deformability index) for concrete reinforced with GFRP reinforcement.

Although GFRP reinforcing bars do not exhibit "classical" corrosion, many researchers [14,15,16,17] have reported that there is a significant reduction in the tensile capacity of GFRP reinforcement when exposed to aggressive solutions and exposure conditions (e.g., high pH, salt water, high temperature, freeze-thaw cycles and wet/dry cycles). Therefore, the long-term performance of concrete members reinforced with FRP bars under harsh environmental conditions has been recognized as an emerging concern. To address these concerns, many researchers adopted accelerated aging test methods using the bare GFRP bars to evaluate the tensile strength of GFRP bars. Although much research has reported significant loss in the tensile capacity of GFRP bars exposed to concrete pore solution, a field study conducted by Mufti *et al.* [18,19] and Benmokrane and Cousin [20] concluded that GFRP reinforcement is durable when embedded in concrete. Recent findings [21] indicate that the interface of fiber and matrix is prone to damage initiation under high temperature (24 °C). Similar debonding was also reported by Benmokrane *et al.* in 2006 [22]. A significant number of simulated aging tests on GFRP bars have been conducted in the past decades. Long-term performance prediction approaches have also been developed, which provide a good basis for the durability evaluation and design of GFRP bars [23]. Based on these results, the current ACI 440.1R-06 [7] requires using a specific environmental reduction factor, C_E, as a design parameter to consider the reduction in the tensile strength of GFRP in actual structures. The value of C_E is varied depending on the exposure conditions of GFRP-reinforced concrete: for concrete not exposed to earth and weather, the reduction factor is 0.8, and for concrete exposed to earth and weather, the reduction factor is 0.7.

However, the information on the impact of the degraded GFRP reinforcement on the flexural performance of concrete members is limited. Furthermore, *in situ* or realistic environmental conditions (*i.e.*, service load level and aggressive temperature and humidity for GFRP reinforcement) were rarely implemented in the experimental program in the literature. Trejo *et al.* [24] reported experimental tension test data for over one hundred GFRP reinforcement embedded in concrete and exposed to actual environmental conditions (mean annual temperature of 23 °C and average annual precipitation of 1008 mm) for seven years. Based on this study, Gardoni *et al.* [25] developed the time-variant model. Later, time-dependent reliability analysis on a bridge deck was analyzed [26]. However, this research has the limited data of the GFRP reinforcement subjected to a higher mean temperature, representing a more aggressive temperature. In addition, this test method performed tension testing of bar samples extracted from the beam specimens. Therefore, further data are needed, and a more systematic experimental plan is needed to consider the impact of the degradations of reinforcing systems embedded in concrete under harsh environmental conditions.

This research provides valuable information on the flexural performance of concrete beams reinforced with GFRP bars after accelerated environmental aging. The main aim of the study is to evaluate the long-term flexural behavior of GFRP in comparison with steel reinforcement by varying the reinforcement ratio and the exposure duration.

EXPERIMENTAL PROGRAM

Materials

The concrete mix design was: 226 kg/m^3 of Type I/II Portland cement, 990 kg/m^3 of coarse aggregate with a maximum size of 10 mm, 1010 kg/m^3 of sand (fine aggregate), 74 kg/m^3 of Class F Fly ash and 198 kg/m^3 of water. The average slump of the concrete was 71 mm, and the average compressive strength (f'_c) at 28 days was 29.5 MPa. The average modulus of elasticity of concrete was estimated to be 25.6 GPa by the 4700 $\sqrt{f'_c}$ equation of ACI 318-11 [27].

Two types of GFRP bars from two different manufacturers were used for the main flexural reinforcement of the beams. Figure 1 shows the reinforcement types used in this study. As shown in Figure 1a, Type A GFRP reinforcement (GA (Glass fiber bar–Aslan) series in the designation of beam specimens) was made of E-glass fiber (wrapped/ partial sand coating surface, E-glass fiber impregnated in a vinyl-ester resin). As shown in Figure 1b, Type B GFRP reinforcement (GH (Glass fiber bar–High modulus) series in the designation of beam specimens) was made of E-glass formulation using the pultrusion-process (sand-coated surface with a high relative modulus of elasticity, vinyl ester-based resin with E-glass formulation). The Type B bar has a much higher glass fiber content at approximately 80% by weight, while the Type A bar has 70% minimum fiber content per the ASTM D2584 "Standard Test Method for Ignition Loss of Cured Reinforced Resins" [23]. Plain carbon-steel bars were used with a minimum yield strength (f_y) of 415 MPa (Grade 60), as shown in Figure 1c. The mechanical properties of glass fiber-reinforced polymer bars and steel bars, as reported by the manufacturer, are presented in Table 1. As shown in Table 1, a Type B bar exhibits approximately a 47% to 57% higher tensile elastic modulus than a Type A bar.

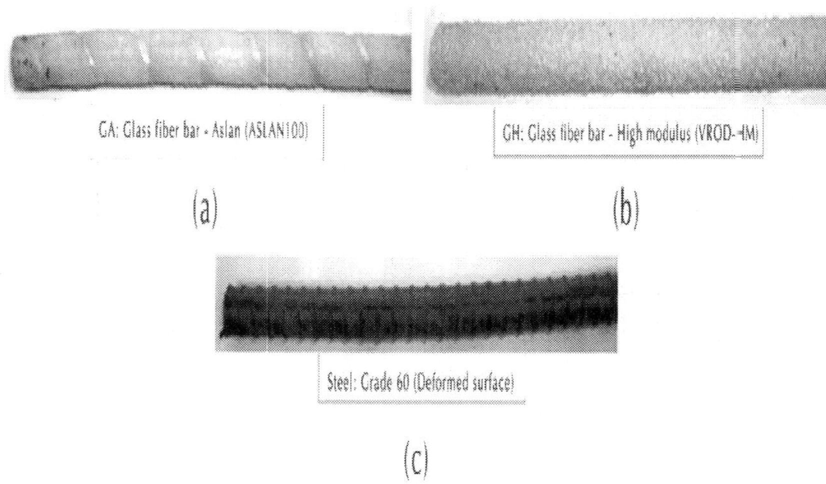

Figure 1. Reinforcements of (**a**) A-type GFRP; (**b**) B-type GFRP and (**c**) steel bars.

Table 1: Mechanical properties of FRP bars

FRP Bar Type	Bar Size	Nominal Diameter (mm)	Nominal Area (mm^2)	Guaranteed Tensile Strength (f_{fu} *: MPa)	Tensile Modulus of Elasticity (E_f *: GPa)	Tensile Strain ($_u$ *: %)
GA series (Type A)	#4	13	126.7	690	40.8	1.50
	#5	16	197.9	655	40.8	1.50
GH series (Type B)	#4	13	126.7	1300	60.0	2.42
	#5	16	197.9	1259	64.1	2.24
Steel (Grade 60)	#4	12.7	129.0	620 (415: yield)	200	9.00
	#5	15.875	200.0	620 (415: yield)	200	9.00

* Test results were obtained in accordance with ACI 440.1R-06 [7]. Grade 60 in the U.S. is equivalent to Grade 420.

Test Matrix and Design of Specimens for Fabrication

A total of thirty six specimens were fabricated in the structural laboratory at the University of Texas at Arlington. The dimensions of each specimen are 200 mm by 300 mm by 1800 mm (width × height × length). For the tensile reinforcement, the longitudinal reinforcement consisted of two #4 (13-mm diameter) or three #5 (16-mm diameter) bars for each specimen with a single layer of reinforcement. The three variables are: (1) the types of reinforcement (Type A and Type B GFRP bars and steel); (2) the reinforcement areas (two #4 or three #5); and (3) the environmental aging (non-exposure and exposure to the designated temperature and humidity for 300 days). Based on the test variables, the nomenclature of test specimens is as follow. The first character, "GA" or "GH" or "S", represents the type of reinforcement for glass fiber of Type A, Type B (relatively higher Young's modulus) and steel, respectively. The second and third characters, "2-4" or "3-5", are the number and size of the reinforcement or varying the reinforcement ratio. For example, 3-5 represents that the reinforcement consists of three #5 bars. The last character, varying from 0 to 300, represents the exposure duration (days).

Failure modes are varied depending on the GFRP reinforcement ratios. According to ACI 440.1R-06 [7], the concrete crushing failure mode is obtained when the reinforcement ratio ($\bullet_f$) is greater than the balanced reinforcement ratio ($\bullet_{fb}$) in Equations (1) and (2):

$$\rho_f = \frac{A_f}{bd} \tag{1}$$

$$\rho_{fb} = 0.85\beta_1 \frac{f'_c}{f_{fu}} \frac{E_f \varepsilon_{cu}}{E_f \varepsilon_{cu} + f_{fu}} \tag{2}$$

where A_f is the area of GFRP reinforcement, b is the width of the section, d is the effective depth, $\bullet_1$ is the ratio of the depth of the equivalent rectangular stress block to the depth of the neutral axis, f'_c is the concrete compressive strength, f_{fu} is the rebar tensile strength, E_f is the modulus of elasticity of the FRP rebar and $\bullet_{cu}$ is the maximum concrete strain (0.003 for ACI provisions).

Figure 2 shows that variations in the reinforcement ratio affect the strength reduction factor in accordance with ACI 440.1R-06 [7]. This chart is used to determine the reinforcement ratio of specimens. It can be seen that the values of $\bullet_f/\bullet_{fb}$ are 0.997 and 4.761 to induce the failure mode of theoretically balanced failure (FRP-rupture and concrete crushing simultaneously: strains of the top fiber in concrete and GFRP bars reach their predefined limiting or useable concrete strain value, $\bullet_{cu}$, and rupture strain of GFRP reinforcement, $\bullet_{fu}$, simultaneously) for GA-2-4-300 and concrete crushing for GH-3-5-0, respectively. Unaged specimens with GFRP reinforcement were designed to fail by concrete crushing, which means that failure was expected to occur when the concrete reached its maximum useable compressive strain ($\bullet_{cu}$). The GA2-4-0 and GA2-4-300 specimens were expected to have a balanced failure, resulting in GFRP rupture and concrete failure. The reinforcement ratios ($\bullet_s/\bullet_{sb}$) of steel are less than those ($\bullet_f/\bullet_{fb}$) of GFRP specimens with the same amount of reinforcement area (two #4 and three #5). It should be noted that this reinforcement ratio led to yielding of the steel reinforcement (referred to as under-reinforced concrete beam) before concrete crushing failure. No stirrups were used in the testing region to secure the pure bending behavior, while #3 steel stirrups were used with a spacing of 100 mm to prevent shear failure at both ends of the beams. The detailed reinforcement layout is shown in Figure 4.

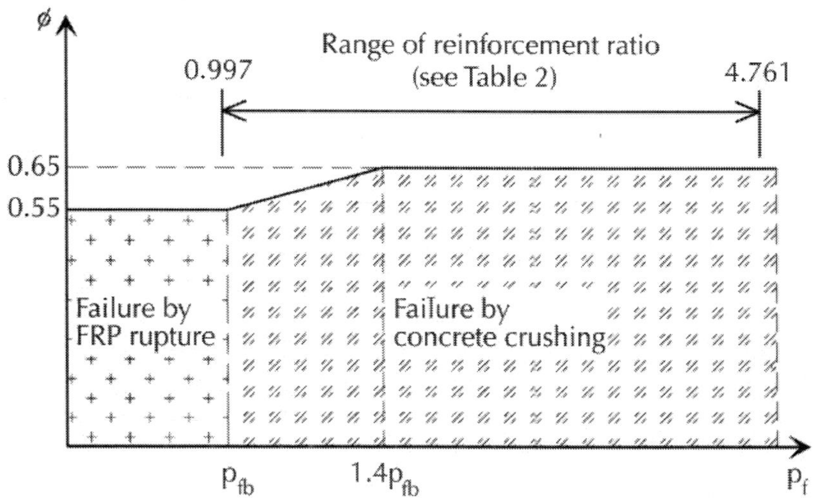

Figure 2: The strength reduction factor as a function of the reinforcement ratio (adapted from ACI 440.1R-06 [7]).

Sustained Loading and Accelerated Aging Method in the Chamber

All beams were clamped in pairs using externally transverse steel rods at the beam end to simulate crack damage under service conditions by a maximum mid-span deflection of *L*/240 (*L*, the clear span of the beam), as shown in Figure 3a. Two roller supports were placed on the top surface of each beam at the one-third location of the beam length measured from each end. The region between two roller supports is considered as a pure bending zone (zero shear zone). Two main pre-cracks appeared in the pure bending zone and were measured by a portable microscope (×200). As shown in Figure 3b, the widths of pre-cracks ranged from 0.26 to 0.61 mm (Average: 0.43 mm), and the average spacing of the pre-cracks was 120 mm (two #4 series) and 164 mm (three #5 series). These cracks were generated at the same 30% of ultimate moment capacity, M_u (*i.e.*, service load level), as the design requirement. This is more valuable information for structural engineers to design and evaluate the performance of structures. Therefore, the differences of the initial damages (for example, between 0.26 mm and 0.61 mm of the width of the pre-crack) were not focused on in this paper. More information on crack widths is reported in Park [28]. Prior to exposure in the chamber, pre-cracked specimens were initially immersed in a 3.0% saline solution for 72 h, and equal end moments were applied to the test beams using clamping. The sustained load level was selected as 0.3*P* (30% of *P*), where *P* is the ultimate flexural load capacity of the beam reinforced with GFRP bars. A constant deformation was maintained during the test through clamping.

For more aggressive environmental conditions, the test focused on the simulated environmental conditions at an average high temperature of 47 °C and an average relative humidity (RH) of 80% based on records from the last 30 years in North Texas in the United States. Re-clamping to simulate service loading in the chamber after immersion into the 3.0% saline solution and placing the specimens in the chamber are shown in Figure 3c. Cracks resulting from the clamping at both ends to simulate service loading were marked, and the crack width and spacing were measured within the constant moment zone. Cracks induced by sustained loading also accelerate the aging effect in the chamber, as shown in Figure 3e.

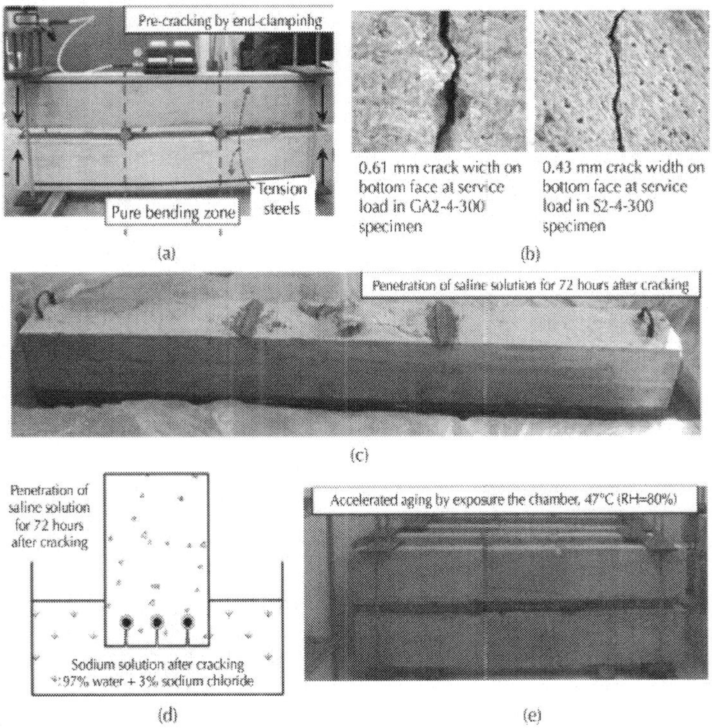

Figure 3: Test procedure: (**a**) pre-cracking for simulating service loading; (**b**) crack widths due to service loading; (**c**) beamimmersed in saline solution; **(d)** penetration of saline solution; and (**e**) accelerated aging in the chamber (47°C and a relative humidity (RH) of 80%).

Four-Point Bending Test

All specimens were subjected to four-point bending after the accelerated environmental conditioning. All tests were displacement-controlled at a rate of 0.15 mm/min. Comparative studies were necessary to investigate the various effects of the conditioning. Figure 4 shows the test set-up and geometry of the beam specimens. The ultimate loads are defined as the maximum loads measured by the load cell used with the data acquisition (DAQ) system. A hydraulic jack applied the load to the reinforced concrete beams with GFRP (single-layered reinforcement) and steel bars through a spreader beam. In order to measure the deflection of the tested beam, two transducers (linear variable differential transducers and strain gauge-based transducers) were used in the mid-span section (pure bending region). An optical magnifier with an accuracy of 0.05 mm was used to measure crack widths at the location of the bottom reinforcement.

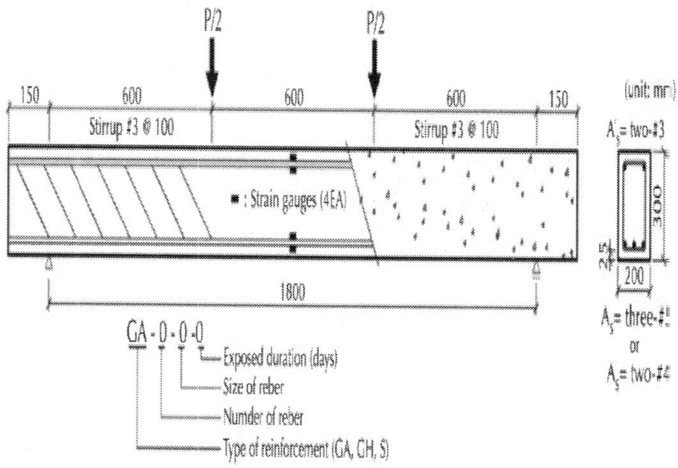

Figure 4: Test set-up and details of a specimen.

TEST RESULTS

This section presents the summary of the overall flexural behavior of the control beams (unaged specimens) and the exposed specimens (aged specimens in accelerated aging for 300 days) in terms of moment-deflection response, failure mode and ductility. The experimental results between control specimens and the exposed specimens for 300 days are also compared and discussed. Finally, predicted deflections and deformability indices of unaged/aged specimens are compared and discussed.

Moment-Deflection Response

Figure 5 shows the experimental moment-deflection curves for concrete beams reinforced with steel reinforcement (RC-steel) and

GFRP reinforcement (RC-GFRP) with varied reinforcement ratios in the mid-span section. Corresponding important values are presented in Table 2. RC-GFRP specimens exhibited no sign of yielding due to the linear-elastic behavior of the GFRP bars until failure, whereas RC-steel specimens with under-reinforcement exhibited yielding of steel before the crushing of concrete. As shown in Figure 5, the specimens, S2-4 series and S3-5 series, exhibited the tri-linear moment-curvature relationship. After the moment exceeded the cracking moment (M_{cr}), regardless of the GFRP bar type and the reinforcement ratio, the flexural stiffness of the beams was significantly reduced, and the curves of RC-GFRP specimens increased almost linearly until the crushing of concrete. The reinforcement ratio, GFRP bar type and the concrete compressive strength all had an influence on the stiffness of the beam specimens. Larger deformations were obtained for lower reinforcement ratios, and *vice versa*. Moreover, for the same reinforcement ratio, GA series (specimens reinforced with Type A GFRP bars) also showed larger deflections due to less relative stiffness compared with the GH series. Except one specimen, GA2-4-300, all specimens demonstrated a concrete crushing mode of failure in line with the design philosophy. The GA2-4-300 specimen exhibited a balanced failure, which indicates the bar failure and concrete crushing failure occurring simultaneously. The moment-carrying capacity was decreased with the occurrence of concrete crushing or balanced failure.

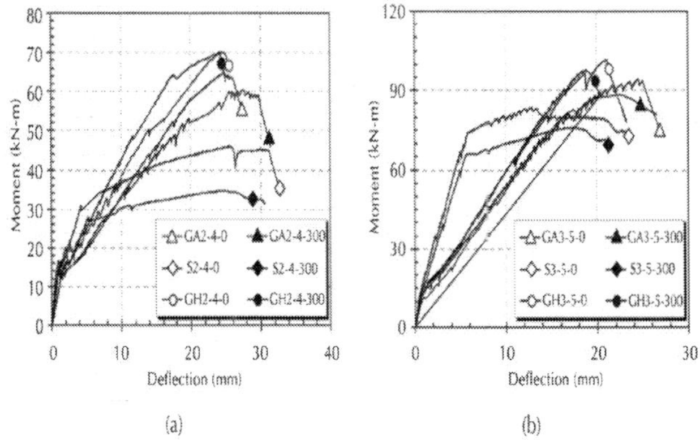

Figure 5: Moment-deflection curves: (**a**) the group of two #4 (GA2-4, GH2-4 and S2-4 series) and (**b**) the group of three #5 (GA3-5, GH3-5 and S3-5 series).

Table 2: Summary of the test results

Specimen I.D.	$_f/_{fb}$ [1] or $_s/_{sb}$ [2]	f'_c (MPa)	$M_{u\text{-}th}$ (kN·m)	$M_{u\text{-}exp}$ (kN·m)				$_{u\text{-}exp}$ (mm)				Failure Mode [3]
				1	2	3	Avg.	1	2	3	Avg.	
GA2-4-0	1.242 [1]	29.30	50.2	64.8	60.9	62.1	62.6	24.49	26.67	22.88	24.68	C.C.
GA2-4-300	0.997 [1]	36.06	51.8	58.6	56.2	55.8	56.9	29.66	30.84	28.43	29.64	B
GA3-5-0	2.157 [1]	29.30	68.1	93.4	88.2	92.1	91.2	17.89	20.81	19.43	19.38	C.C.
GA3-5-300	1.678 [1]	36.47	72.9	86.1	79.4	82.6	82.7	21.31	22.45	23.36	22.37	C.C.
GH2-4-0	2.273 [1]	29.92	52.6	70.1	66.7	68.9	68.6	23.14	25.31	23.76	24.07	C.C.
GH2-4-300	1.776 [1]	36.47	59.9	69.9	68.1	66.9	68.3	28.93	26.43	25.76	27.04	C.C.
GH3-5-0	4.761 [1]	29.92	74.8	101.4	98.3	99.4	99.7	17.22	21.09	20.36	19.56	C.C.
GH3-5-300	3.578 [1]	36.61	85.2	96.1	94.0	–	95.1	20.75	23.41	–	22.08	C.C.
S2-4-0	0.155 [2]	29.10	27.3	45.9	41.7	43.7	43.8	31.17	30.83	33.14	31.71	S.Y. + C.C.
S2-4-300	0.119 [2]	36.06	27.8	34.6	33.9	–	34.3	35.14	35.29	–	35.22	S.Y. + C.C.
S3-5-0	0.352 [2]	29.92	60.1	84.6	82.1	80.6	82.4	25.29	24.80	27.16	25.75	S.Y. + C.C.
S3-5-300	0.267 [2]	36.75	61.8	75.1	72.6	–	73.9	21.34	20.09	–	20.72	S.Y. + C.C.

Notes: f'_c is the average concrete strength at testing day, but those at 28 days were 29.30 MPa in the GA series, 29.92 MPa in the GH series and 29.51 MPa in the S series; $M_{u\text{-}th}$ is the theoretical ultimate moment; and $M_{u\text{-}exp}$ and $_{u\text{-}exp}$ are the experimental ultimate moment and its corresponding deflection. [1] $_f/_{fb}$ (see Equations (1) and (2)); [2] $_s/_{sb}$, where $_s = A_s/bd$ and $_{sb} = (0.85\ _1\ f'_c/f_y)(0.003/(0.003+\ _y))$; [3] C.C., concrete crushing; B, balanced failure; S.Y. + C.C., steel yielding and concrete crushing; –, no data available; Avg., Average.

In all the cases, except the GA2-4 series, however, the moment-carrying capacity decreased and the deflection increased as a function of time when exposed to accelerated aging conditions. In general, all beams showed a reduction in moment-carrying capacity after 300 days of accelerated aging in the environmental chamber (with 47 °C and relative humidity of 80%). Environmental conditioning had an effect on the structural degradation of not only RC-steel, but also RC-GFRP specimens. In all of the cases, the reductions in moment-carrying capacity were observed under the aggressive aging conditions to which the specimens were exposed. RC-steel specimens had a higher reduction in flexural strength than RC-GFRP specimens in both reinforcement cases, as expected in 21.7% and 10.3% in the S2-4 and S2-5 series, respectively. The GA2-4, GH2-4 and S2-4 series exhibited a reduction in the moment-carrying capacity of approximately 9.1%, 0.5% and 21.7%, respectively. Furthermore, the GA3-5, GH3-5 and S3-5 series exhibited a reduction in the moment-carrying capacity of approximately 9.3%, 4.6% and 10.3%, respectively, after exposure for 300 days in the environmental chamber. The substitution of GFRP reinforcement for steel reinforcement in the flexural members is beneficial for improving the durability of reinforced beams. From the results, the rate of flexural strength degradation of Type A GFRP beams (*i.e.*, GA series) is greater than that of Type B GFRP beams (*i.e.*, GH series). This can be attributed to the coating materials of GFRP reinforcement. The Type B GFRP bars in the GH series with a sand-coated surface exhibited better protection against moisture and temperature as compared to the Type A GFRP bars in the GA series. It should be noted that among the entire specimens, the lowest reduction was observed in the GH2-4-0 and GH2-4-300 specimens. A higher GFRP reinforcement with larger-sized diameter bars led to a decrease in the rate of reduction in the flexural strength (GA2-4 series *versus*GA3-5 and GH3-5 series). Whereas the observed corrosion of the steel reinforcement led to significant degradation of the flexural strength, the GFRP reinforcement showed no significant changes (below 10%) in the flexural strength. Within this range, only marginal degradation of GFRP reinforcement occurred.

The theoretical ultimate moments ($M_{u\text{-}th}$) were calculated according to ACI 440.1R-06 [7] with the ultimate concrete strain ($_{cu}$) of 0.003. Concrete compressive strengths were considered when calculating reinforcement ratios tabulated in Table 2. All beams exhibited a higher capacity than the theoretically predicted value. Mean ratios of RC-GFRP beams between the experimental and theoretical values calculated by ACI 440.1R-06 [7] are 1.20 and 1.22, respectively, before and after the aging. The ratio above 1.0 represents that the design approach of ACI 440.1R-06 is still conservative and appropriate.

Deflection

Using various existing models, the calculated deflections at mid-span for a simply-supported beam of total length L subjected to four-point bending are given in Table 3. As mentioned earlier, the low elastic modulus of GFRP reinforcement requires the design of RC-GFRP specimens at the serviceability limit state in accordance with ACI 440.1R-06 [7]. Therefore, an assessment is needed to understand the influence of the low stiffness of GFRP reinforcement coupled with degradation on overall flexural deflection.

The ACI 440.1R-06 model [7] recommends using Equations (3) and (4) to calculate the effective moment of inertia (I_e) (see Table 3). The factor, $_d$, is a reduction factor related to the reduced tension stiffening exhibited by GFRP-reinforced members [8]. The approach of Toutanji and Saafi [29] of Equation (5) was also compared with experimental data. In Toutanji and Saafi's model, the exponent of Branson's model [8] was modified by adopting E_s of the steel modulus of elasticity, $_f$ of the FRP reinforcement ratio and E_f of the FRP modulus of elasticity when using this method. Finally, Bischoff and Gross's model [30] of Equation (6) was also compared to the experimental data. As seen in Equation (6), an equivalent moment of inertia based on the change in stiffness along the span was adopted in Bischoff and Gross's expression.

Table 3: The equivalent moment of inertia (I_e) for the calculation of deflections under the four-point bending

Loading and support condition	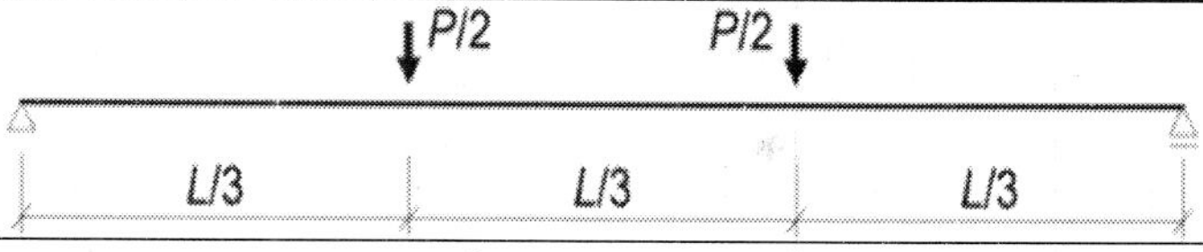
Theoretical deflection	$$\Delta_{mind-span} = \frac{23P^3}{1296E_cI_e} \quad (3)$$
ACI 440 1.R-06 [7]	$$I_e = \left(\frac{M_{cr}}{M_a}\right)^3 \beta_d I_g + \left[1-\left(\frac{M_{cr}}{M_a}\right)^3\right] I_{cr} \le I_g \quad (4)$$
Toutanji and Saffi [29]	For $\left(\frac{E_f}{E_s}\right)\rho_f < 0.3$, $$I_e = \left(\frac{M_{cr}}{M_a}\right)^{6-\frac{10\rho_f E_f}{E_s}} I_g + \left[1-\left(\frac{M_{cr}}{M_a}\right)^{6-\frac{10\rho_f E_f}{E_s}}\right] I_{cr} \le I_g \quad (5a)$$ $$I_e = \left(\frac{M_{cr}}{M_a}\right)^3 I_g + \left[1-\left(\frac{M_{cr}}{M_a}\right)^3\right] I_{cr} \le I_g \quad (5b)$$ For $\left(\frac{E_f}{E_s}\right)\rho_f \ge 0.3$,
Bischoff and Gross [30]	$$I_e = \left(\frac{I_{cr}}{1-\gamma\eta\left(\frac{M_{cr}}{M_a}\right)^2}\right) \quad (6)$$ where $M_a = \frac{PL}{6}, \eta = 1-\frac{I_{cr}}{I_g}$ For considering tension stiffening, $\gamma = 1.72 - 0.72\frac{M_{cr}}{M_a}$ For considering no tension stiffening in the cracked region, $\gamma = 0.35\frac{M_{cr}}{M_a}$

Notes: E_c = concrete modulus of elasticity; E_s = steel modulus of elasticity; I_g = moment of inertia of gross section; M_a = applied moment ($PL/6$); all other parameters are presented in the text.

As compared with theoretical predictions obtained using ACI 440.1R-06 [7] and other existing models, Figure 6 shows the experimental data and predicted values of deflections for 300-day aged and control RC-GFRP specimens subjected to four-point loading (only the GA2-4 series). Note that samples that exhibited the maximum capacities in each case are compared. Similar trends are observed for all other specimens. After 300 days of accelerated aging in the chamber, the deflections of the GA2-4 (or GH2-4) and GA3-5 (or GH3-5) beams reinforced with Type A GFRP bars experienced an average increase of approximately 21% (or 25%) and 19% (or 20%), respectively, when comparing GA2-4-0 (or GH2-4-0) with GA2-4-300 (or GH3-5-300). This indicates that the accelerated aging exaggerated the effect of the smaller amount of reinforcement (or stiffness) on the increase in deflection (e.g., GA2-4 *versus* GA3-5; GH2-4 *versus* GH3-5). The larger deflection after aging may be due to the degradation in the cracked stiffness (E_cI_{cr}) of specimens. Up to the cracking moment in flexural tests, there were no significant differences between any of the deflections and cracking moments for RC-GFRP beams. However, from approximately 30% of the ultimate moment (corresponding to the service level) up to failure, the theoretical approaches showed different trends in Figure 6.

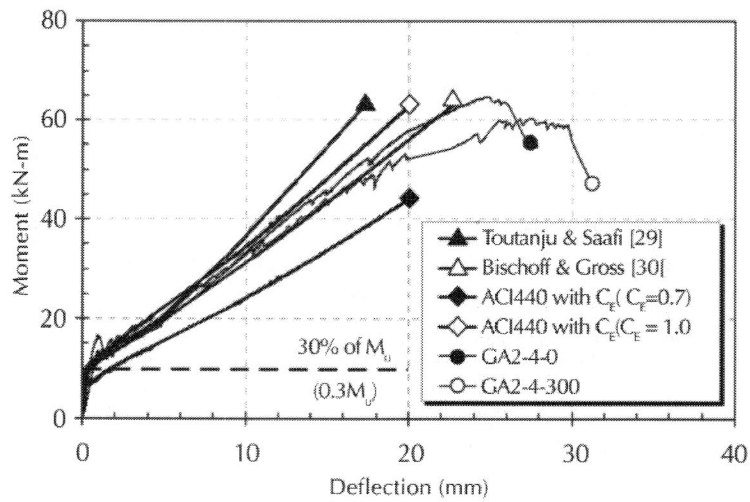

Figure 6: Comparison of predicted and measured results in the GA2-4 series.

The Bischoff and Gross model [30] overestimated deflection, while ACI 440.1R-06 [7] without C_E (*i.e.*, C_E = 1.0) and the Toutanji and Saafi model [29] underestimated deflection in all cases at the service (0.3 M_u) and ultimate load level, M_u. Differences between theoretical and experimental data were not significant, as the reinforcing ratio increased, because tension stiffening (simply, defined as tension force carried by the concrete between the two adjacent or multiple cracks) decreased at the higher reinforcing ratios. When an environmental reduction factor of 0.7 (C_E: exposed to earth and weather in ACI 440.1R-06 [7]) was used to calculate the deflection of a member with concrete crushing failure, the ACI 440 provision [7] was still conservative for specimens subjected to the 300 days of accelerated aging. Accordingly, the equations underestimated the deflection at the ultimate moment. Similarly, the prediction models for all cases tend to overestimate the value of I_e, resulting in the underestimation of the deflection. The exposure condition exaggerates the overestimation of prediction models on the deflection at the ultimate moment. The actual loss of stiffness over time in this specimen did not occur as much as the predicted values based on C_E = 0.7 in ACI 440.1R-06 [7]. Therefore, the accuracy of the predicted deflections gets lower as the load increases.

Table 4 shows the comparisons between experimental and predicted deflections at the ultimate moment. Note that samples that exhibited the maximum moment capacities in each case are compared. Similar trends are observed for all other specimens. The ratio of predicted value to experimental value for the Bischoff and Gross equation [30] exhibited a mean value of 0.99 (0.82) and standard deviations of 0.09 (0.09) for unaged specimens (aged specimens). The Bischoff and Gross equation [30] exhibited the smallest average error for the long-term performance of aged specimens. Thus, it provided the best predictions of load and deflections for the accelerated aged specimens, while the Toutanji and Saafi model [29] exhibited poor performance on predicting the deflection in not only the accelerated aged specimens, but also control specimens. From these results, Toutanji and Saafi model [29] exhibited an inappropriate model for the design, because it underestimated short-term and long-term deflections considerably for GFRP-reinforced concrete beams.

Table 4: Deflections at the ultimate moment

Specimen I.D.		Reinforcement Ratio (A_f/ bd) Exp. *	Deflection (mm) Eq. (4) (C_E = 1.0)	Eq. (5)	Eq. (6)		$\frac{\text{Eq. (4)}}{\text{Exp.}}$	$\frac{\text{Eq. (5)}}{\text{Exp.}}$	$\frac{\text{Eq. (6)}}{\text{Exp.}}$
GA2-4-0		0.0045 (0.45%)	24.488	21.699	16.969	23.404	0.886	0.693	0.956
GA2-4-300			29.655	21.699	16.969	23.404	0.732	0.572	0.789
GA3-5-0		0.0109 (1.09%)	17.882	16.318	13.257	19.870	0.913	0.741	1.111
GA3-5-300			21.311	16.318	13.257	19.870	0.766	0.622	0.932
GH2-4-0		0.0045 (0.45%)	23.146	17.672	14.552	20.359	0.764	0.629	0.880
GH2-4-300			28.931	17.672	14.552	20.359	0.611	0.503	0.704
GH3-5-0		0.0109 (1.09%)	17.223	15.160	12.866	17.509	0.880	0.747	1.017
GH3-5-300			20.752	15.160	12.866	17.509	0.731	0.620	0.844
0 day	Mean	–	–	–	–	–	0.86	0.71	0.99
	Std.	–	–	–	–	–	0.07	0.06	0.09
300 days	Mean	–	–	–	–	–	0.71	0.58	0.82
	Std.	–	–	–	–	–	0.07	0.06	0.09

Notes: Eq. means Equation; Std. means Standard Deviation; Equation (4) is taken in ACI 440-1R-06 [7]; Equation (5) is taken in Toutanji and Saafi [29]; and Equation (6) is taken in Bischoff and Gross [30]. * Exp., experimental deflection at mid-span at the ultimate stage of one sample in each case.

Failure Mode and Cracks

When the cracking moment was reached in the pure bending zone, some cracks began to appear, and the existing cracks were propagated and opened. These cracks were predominantly vertical and perpendicular to the direction of the maximum stress induced by the bending moment.

As the load increased, additional cracks developed in the mid-span and new vertical cracks formed in the shear span. The cracks that developed in the tested specimens of RC-GFRP and RC-steel specimens at the ultimate stage are shown in Figure 7. For RC-GFRP beams, cracking initiated in the constant moment region, with the cracks originating from the bottom fibers, as the principal stresses were the greatest at these extreme fibers. These initial cracks traversed quite deeply into the compression zone. The height of the initial cracks was varied from 64 to 99 mm, measured from the bottom fiber. The crack spacing also decreased rapidly with increasing load. This is also observed in the specimens exposed for 300 days. However, the spacing of cracks remained approximately constant after the applied moment of approximately 46 kN·m for all the cases. Beyond this load level, the existing cracks propagate into the compression zone without any newly forming cracks. Similar to RC-GFRP beams, cracking in RC-steel beams also started with vertical flexural cracks in the constant moment region. However, for RC-steel beams, very few cracks outside the pure bending zone turned into inclined cracks, and most of these were relatively vertical, as shown in Figure 7c. These beams showed a classic reinforced concrete crack pattern involving fewer and larger cracks. The spacing of cracks, however, decreased as the load increased.

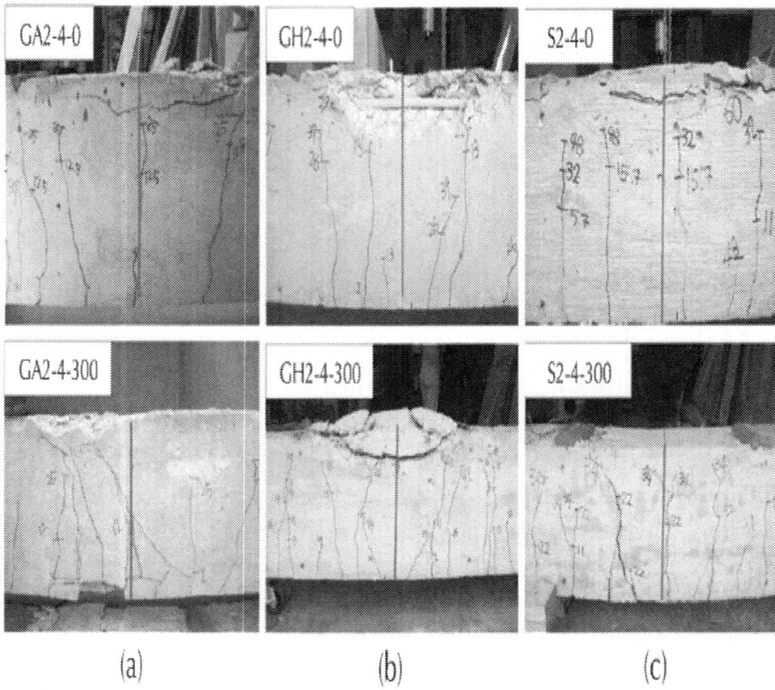

Figure 7: Crack pattern at the ultimate stage of the (**a**) GA2-4-0 and GA2-4-300; (**b**) GH2-4-0 and GH2-4-300; and (**c**) S2-4-0 and S2-4-300 specimens.

Ductility

The conventional concept of ductility is based on the beam's capacity to absorb energy without critical failure and is generally related to the post-yielding deformational response before bar failure or concrete crushing failure. However, the concept of ductility in its traditional consensus in RC-steel is not applicable for RC-GFRP beams due to non-existent yielding or no post-yielding deformation of GFRP reinforcement. It is for this reason that many researchers and design codes have suggested other ductility models referred to as pseudo-ductility measures. In order to discuss the flexural ductility of RC-GFRP specimens quantitatively, the term deformability has been introduced by researchers as a means of assessment of the displacement, curvature, applied moment and energy that occurs before the rupture of the GFRP and other types of FRP reinforcement [31,32,33,34,35]. Researchers have proposed several different models to evaluate the deformability of concrete beams with FRP bars, including GFRP reinforcement. However, there is still a lack of general agreement as to how the deformability characteristics of such members may be quantified and analyzed, even for unaged specimens.

Table 5 shows the parameters of each specimen for estimating ductility indices. Note that samples that exhibited the maximum moment capacities in each case are compared. Similar trends are observed for all other specimens. They are curvature, deflection and moment, at different stages. The definition of each parameter is presented in Table 5. Naaman and Jeong [33] proposed an energy-based deformability index (referred to as the Naaman Index) to compute the ductility index of μ_e for beams reinforced or prestressed with GFRP tendons. This index is suitable for beams with steel reinforcement, as well as GFRP reinforcement [31]. Grace *et al.* [31] proposed the equation as a function of four parameters. The parameters are: the influence of the type of reinforcement (e.g., longitudinal reinforcement and stirrups), ; the failure mode of the beam, ; the ratio of modulus of elasticity of GFRP to steel reinforcement, E_f/E_s; and the failure strength of the reinforcement, . These parameters are used to determine the slope of the load-deflection as the parameter of *S* in Equation (8) (referred to as the Grace Index).

Table 5: Deformability indices

Approach Type	**Deformability Index**	
Energy-based approach 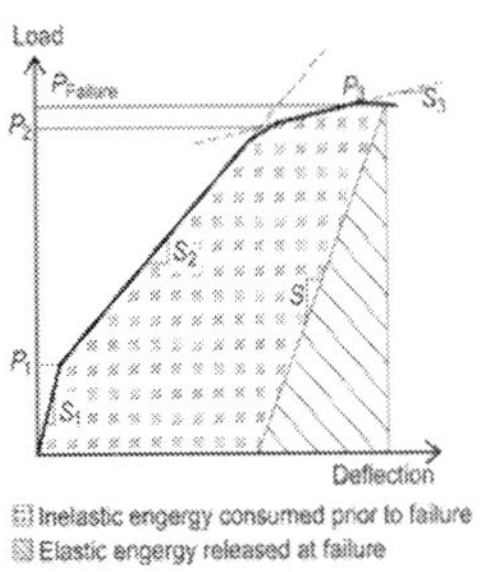	Naaman and Jeong [33]	$\mu_e = 0.5(E_T / E_{el} + 1)$ (7a) $S = [P_1S_1 + (P_2 - P_1)S_2] / P_2$ (7b)
	Grace *et al.* [31]	$S = \alpha\beta\gamma \frac{E_f}{E_s}\frac{f_y}{f_{ds}}\frac{P_1S_1 + (P_2 - P_1)S_2 + (P_3 - P_2)S_3}{P_3}$ (8) : stirrup factor (GFRP: 0.95) : failure mode factor (compression and flexure: 1.0) : reinforcement factor (GFRP: 4.0)

Moment and deformation-based approach	Jaeger *et al.* [32]	$$\mu_E = \frac{\varphi_u}{\varphi_{0.001}} \frac{M_u}{M_{0.001}} \quad (9)$$
	Zou [34]	$$Z = \mu_E = \frac{\Delta_u}{\Delta_{cr}} \frac{M_u}{M_{cr}} \quad (10)$$
Deflection-based approach 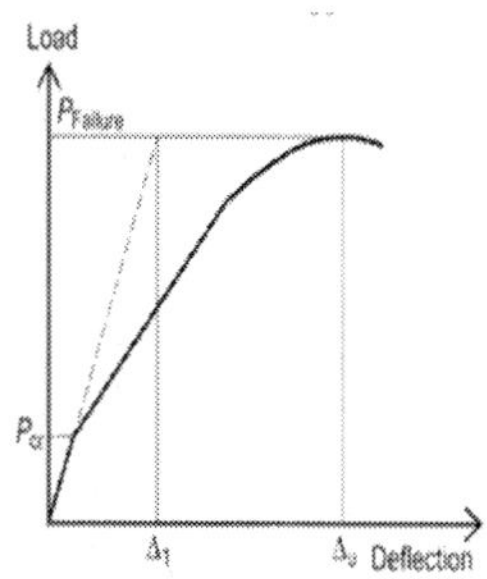	Abdelrahman *et al.* [35]	$$A_b = \frac{\Delta_u}{\Delta_1} \quad (11)$$

Notes: P_1 and P_2 = loads as shown in the energy-based approach; S_1 and S_2 = corresponding slope; E_T = the total energy as the area under the load-deflection curve up to the failure load; E_{el} = elastic energy released at the failure; f_{ds} = design strength of FRP; E_f = FRP modulus of elasticity; $_u$ = deflection corresponding to the ultimate load; and $_{cr}$ = deflection at cracking loads.

Jaeger *et al.* [32] defined the service limit state as the strain state corresponding to a maximum compressive strain in concrete of 0.001 and used the moment and curvature ratios at the ultimate and service limit states (referred to as the Jaeger Index). The Intelligent Sensing for Innovative Structures (ISIS-Canada) [36] adopted the Jaeger model, with deformability index values greater than four and six for rectangular sections and T-sections, respectively. Zou [34] proposed a deformability index (referred to as the Zou Index) defined in terms of both a moment factor and a deflection factor for FRP prestressed concrete beams. The moment part is the ratio of the ultimate moment to the cracking moment, while the deflection part is the ratio of the deflection at failure to that at first crack. Abdelrahman *et al.* [35] established a deformability model based on deflection for beams prestressed by FRP tendons (referred to as the Abdelrahman Index) [35]. The value of A_b, is the ratio of the maximum deflection, Δ_u corresponding to the failure or maximum load to the equivalent deflection, Δ_1, of the uncracked section at a load equal to the ultimate load. It should be noted that the Naaman [33], Zou [34] and Abdelrahman indices [35] have been developed for FRP prestressed concrete beams defined in terms of both a deflection factor and a moment factor.

Although there are different ways to calculate the ductility index, there is no doubt that ductility is the ability to absorb inelastic energy without losing load-carrying capacity. From this standpoint, it is necessary to evaluate the change of deformability after accelerated aging for long-term performance.

Table 7 summarizes the variation of deformability indices with different exposure conditions and durations (*i.e.*, control *versus* 300 days) for the six groups of beams studied based on the test results in Table 6. Note that samples that exhibited the maximum moment capacities in each case are compared. Similar trends are observed for all other specimens. Higher amounts of reinforcement (*i.e.*, GA3-5 and GH3-5 series) reduced the deformation and curvatures, resulting in the reduction of the value of the deformability indices (*i.e.*, GA2-4 and GH2-4 series) (see Table 7). For example, it is obvious that RC-GFRP beams have lower values of Δ_u/Δ_{cr} than RC-steel beams. Furthermore, compression failure is observed in the specimens containing higher amounts of reinforcement (*i.e.*, GA3-5 and GH3-5 series). This can be attributed to plastic hinge formation and significant concrete cracking in the compression zone, as well as stress redistribution.

All ductility indices showed inconsistent compatibility with the deformability of aged specimens except the Jaeger index [32], which is consistent in all cases. For all cases, the normalized Jaeger index decreased for aged specimens as compared to unaged specimens. It indicates that the Jaeger index is appropriate for comparing the ductility of aged and unaged specimens. However, the other indices do not exhibit the consistent trend in some cases. For example, the Abdelrahman index [35] increased from 6.03 (GA2-4-0) to 6.14 (GA2-4-300) after the aged condition.

Table 6: Experimental bending moments, displacements and curvatures of specimens at cracking, service and ultimate stages for estimating the ductility indices in Table 7

Specimen I.D.	$M_{0.001}$ (kN·m)	M_{cr} (kN·m)	M_u (kN·m)	$_{0.001}$ (rad/mm)	$_u$ (rad/mm)	$_{cr}$ (mm)	$_u$ (mm)	$_u/_{cr}$	$_1$ (mm)
GA2-4-0	22.07	16.10	64.82	0.000024	0.000048	1.245	24.488	19.67	4.064
GA2-4-300	24.23	15.23	58.57	0.000031	0.000071	1.372	29.655	21.61	4.826
GA3-5-0	29.65	16.03	93.40	0.000016	0.000054	0.737	17.881	24.26	3.302
GA3-5-300	27.19	15.58	86.07	0.000018	0.000059	0.991	21.311	21.51	4.572
GH2-4-0	25.47	14.68	70.12	0.000016	0.000059	1.194	23.146	19.38	4.572
GH2-4-300	22.24	14.26	69.85	0.000021	0.000067	1.092	28.932	26.49	5.334
GH3-5-0	33.00	16.33	101.39	0.000013	0.000044	0.533	17.223	32.31	3.302
GH3-5-300	27.84	15.70	96.07	0.000016	0.000048	0.787	20.752	26.37	5.334
S2-4-0	31.21	15.01	45.92	0.000010	0.000079	0.787	31.173	39.61	2.667
S2-4-300	28.74	12.02	34.60	0.000012	0.000096	0.533	35.143	65.93	2.159
S3-5-0	34.03	15.13	84.62	0.000008	0.000071	1.143	25.287	22.12	3.175
S3-5-300	32.56	12.02	75.05	0.000009	0.000083	0.787	21.343	27.12	2.413

Notes: $M_{0.001}$ ($_{0.001}$) = moment (curvature) at a service limit corresponding concrete strain (compression zone) of 0.001; M_{cr} = cracking moment; M_u = ultimate moment; $_u$ = curvature at ultimate state; $_{cr}$ = deflection at cracking loads; and $_1$ = the equivalent deflection of the uncracked section at a load equal to the ultimate load).

Table 7: Comparison of deformability indices and normalized deformability

Specimen I.D.	Deformability Index					Normalized Deformability				
	Naaman [33]	**Grace [31]**	**Zou [34]**	**Jaeger [32]**	**Abdelrahman [35]**	**Naaman [33]**	**Grace [31]**	**Zou [34]**	**Jaeger [32]**	**Abdelrahman [35]**
GA2-4-0	2.13	1.31	79.19	5.87	6.03	1.00	1.00	1.00	1.00	1.00
GA2-4-300	1.51	0.99	83.12	5.54	6.14	0.71	0.76	1.04	0.93	1.02
GA3-5-0	1.12	0.98	141.36	10.63	5.42	1.00	1.00	1.00	1.00	1.00
GA3-5-300	0.98	1.11	118.80	10.38	4.66	0.88	1.13	0.84	0.98	0.86
GH2-4-0	1.35	1.13	92.59	10.15	5.07	1.00	1.00	1.00	1.00	1.00
GH2-4-300	1.04	1.29	129.78	10.02	5.42	0.77	1.14	1.40	0.99	1.07
GH3-5-0	1.28	1.11	200.63	10.40	5.22	1.00	1.00	1.00	1.00	1.00
GH3-5-300	1.32	1.03	161.35	10.25	3.89	1.03	0.93	0.80	0.99	0.75
S2-4-0	6.22	4.16	121.18	11.62	11.63	1.00	1.00	1.00	1.00	1.00
S2-4-300	7.51	5.69	189.78	9.63	16.28	1.21	1.37	1.57	0.83	1.40
S3-5-0	2.93	1.96	123.73	22.07	7.96	1.00	1.00	1.00	1.00	1.00
S3-5-300	4.57	3.06	169.33	21.26	8.85	1.56	1.56	1.37	0.96	1.11

As mentioned earlier, the inconsistent results may be attributed to the fact that the Naaman [33], Zou [34] and Abdelrahman indices [35] have been developed for FRP prestressed concrete beams. Furthermore, the inelastic slope (S_2), the cracking deflection (Δ_{cr}) and the equivalent deflection (Δ_1) are difficult to assess experimentally, due to the linear elastic behavior of GFRP material up to the rupture of the GFRP reinforcement without the sign of yielding that usually exists in RC-steel. Moreover, the deformability indices of the Jaeger index [32] were more than two times higher than the recommended minimum value of four of the ISIS Canada design manual. Comparison of the deformability for RC-GFRP beams and RC-steel beams indicates that the values are not similar, since RC-steel specimens showed a true yielding behavior. RC-steel specimens exhibited an increase of the deformability for aged specimens, except for when the Jaeger index [32] is used.

However, it is clear that there is still no general agreement on the quantification of an appropriate index for aged specimens. Essentially, more data are needed. Further investigation is necessary to determine the requirements for deformability.

CONCLUSIONS

This study presents the flexural behaviors and ductility of concrete beams reinforced with GFRP bars after accelerated environmental aging. A total of thirty six beams, with different reinforcement amounts of GFRP and different types of reinforcement, were subjected to sustained loads and placed for 300 days in an environmental chamber. The following concluding remarks have been developed:

1. The substitution of GFRP reinforcement for steel bars in the concrete beams affected the load-deflection response with higher strength and less deflection at the failure stage due to the non-ductile behavior of FRP reinforcement. Flexural GFRP reinforcement in the concrete led to a significant improvement in the flexural strength.
2. Environmental conditioning has an effect on the flexural strength degradation of both RC-steel and RC-GFRP beams. The test results confirm that steel reinforcement had a higher reduction in flexural strength than GFRP reinforcement. It can be concluded that the

substitution of GFRP reinforcement for steel reinforcement in the flexural members affected the durability positively.

3. The rate of flexural strength degradation of concrete beams reinforced with Type A GFRP reinforcement (*i.e.*, GA series) was greater than those with Type B GFRP reinforcement (*i.e.*, GH series). This can be attributed to the coating materials: the Type B GFRP bars with a sand-coated surface seem to have better protection against moisture and temperature.
4. For both unaged and aged specimens, the predictions of flexural deflection by the ACI 440.1R-06 [7], Toutanji and Saafi [15] and Bischoff and Gross [30] models were in close agreement with the experimental data at the service load level. However, as the applied load and aging duration increased, the ACI 440.1R-06 [7] and Toutanji and Saafi [29] models tend to underestimate deflections. The Bischoff and Gross model [30] provided the smallest error and is the best predictor of deflections before and after accelerated aging.
5. If an environmental strength reduction factor (C_E) of 0.7 was used for the flexure design of a member failing in compression with GFRP reinforcement, the ACI 440.1R-06 model [7] was still conservative under the accelerated environmental aging, with a 47 °C temperature and 80% relative humidity. Therefore, the C_E factor is appropriate for estimating the reduced capacity of GFRP reinforcement until 300 days of accelerated environmental aging.
6. Except for the Jaeger index [32], there is no general trend related to the aging duration (days). Rationally, the Jaeger index [32] represents the degradation of ductility due to the aging conditions. However, the deformability indices attained from the Jaeger index were higher than the recommended value of four in ISIS-Canada. Finally, there is still no general agreement on how much degradation of deformability is enough for the aged structures. Further investigation is necessary to determine the requirement of deformability.

ACKNOWLEDGMENTS

The work described in this paper has been supported by Hughes Brothers Inc. (Seward, NE, USA) and Concrete Protection Product. Inc. (Kernersville, NC, USA).

REFERENCES

1. Almusallam, T.H.; Al-Salloum, Y.A.; Alsayed, S.H.; Mosallam, A.S. Durability and long-term behavior of reinforced concrete beams strengthened with FRP composites. In Proceedings of International Conference on FRP Composites in Civil Engineering (CICE 2001), Hong Kong, China, 12–15 December 2001; pp. 1579–1588.
2. Micelli, F.; Nanni, A. Durability of FRP rods for concrete structures. *Constr. Build. Mater.* 2004, *18*, 491–503.
3. Böer, P.; Holliday, L.; Kang, T.H.-K. Independent environmental effects on durability of fiber-reinforced polymer wraps in civil applications: A review. *Constr. Build. Mater.* 2013, *48*, 360–370.
4. Böer, P.; Holliday, L.; Kang, T.H.-K. Interaction of environmental factors on fiber-reinforced polymer composites and their inspection & maintenance: A review. *Constr. Build. Mater.* 2014, *50*, 209–218.
5. Choi, D.-U.; Kang, T.H.-K.; Ha, S.-S.; Kim, K.-H.; Kim, W. Flexural and bond behavior of concrete beams strengthened with hybrid carbon-glass fiber-reinforced polymer sheets. *ACI Struct. J.* 2011, *108*, 90–98.
6. American Association of State Highway and Transportation Officials (AASHTO). *AASHTO LRFD Bridge Design Guide Specifications for GFRP-Reinforced Concrete Bridge Decks and Traffic Railings*; AASHTO: Washington, DC, USA, 2009; p. 68.
7. American Concrete Institute (ACI). *Guide for the Design and Construction of Structural Concrete Reinforced with FRP Bars*; ACI 440.1R-06. ACI: Farmington Hills, MI, USA, 2006.
8. Branson, D.E.; Metz, G.A. *Instantaneous and Time-Dependent Deflections of Simple and Continuous Reinforced Concrete*

Beams; Alabama Highway Department, Bureau of Public Roads: Montgomery, AL, USA, 1965.

9. Razaqpur, A.G.; Isgor, O.B. Rational method for calculating deflection of continuous FRP R/C beams. *ACI Spec. Publ.* 2003, *210*, 191–208.

10. Gao, D.; Benmokrane, B.; Masmoudi, R. *A Calculating Method of Flexural Properties of FRP-Reinforced Concrete Beam: Part 1–Crack Width and Deflection*; Department of Civil Engineering, University of Sherbrooke: Sherbrooke, QC, Canada, 1998.

11. Au, F.T.K.; Du, J.S. Deformability of concrete beams with unbonded FRP tendons. *Eng. Struct.* 2008, *30*, 3764–3770.

12. Jeong, S. Evaluation of Ductility in Prestressed Concrete Beams Using Fiber Reinforced Plastic Tendons. *Ph.D. Thesis*, University of Michigan, Ann Arbor, MI, USA, January 1994.

13. Robert, M.; Benmokrane, B. Combined effects of saline solution and moist concrete on long-term durability of gfrp reinforcing bars. *Constr. Build. Mater.* 2013, *38*, 274–284.

14. Trejo, D.; Aguiniga, F.; Yuan, R.; James, R.W.; Keating, P.B. *Characterization of Design Parameters for Fiber Reinforced Polymer Composite Reinforced Concrete Systems*; Texas Transportation Institute Research Report. The Texas A&M University System: College Station, TX, USA, 2005.

15. Tannous, F.E.; Saadatmanesh, H. Environmental effects on the mechanical properties of E-glass frp rebars. *ACI Mater. J.* 1998, *95*, 87–100. [Google Scholar]

16. Mukherjee, A.; Arwikar, S.J. Performance of glass fiber-reinforced polymer reinforcing bars in tropical environments—Part I: Structural scale tests. *ACI Struct. J.* 2005, *102*, 745–753.

17. Bakis, C.E.; Nanni, A.; Terosky, J.A.; Koehler, S.W. Self-monitoring, pseudo-ductility, hybrid FRP reinforcement rods for concrete applications. *Compos. Sci. Technol.* 2001, *61*, 815–823.

18. Mufti, A.; Banthia, N.; Benmokrane, B.; Boulfiza, M.; Newhook, J. Durability of GFRP composite rods. *Am. Concr. Inst. Concr. Int.* 2007, 37–42.

19. Mufti, A.; Onofrei, M.; Benmokrane, B.; Banthia, N.; Boulfiza, M.; Newhook, J.P.; Bakht, B.; Tadros, G.S.; Brett, P. Field study of

glass-fibre-reinforced polymer durability in concrete. *Can. J. Civ. Eng.* 2007, *34*, 355–366.

20. Benmokrane, B.; Cousin, P. *University of Sherbrook GFRP Durability Study Report*; Intelligent Sensing for Innovative Structures (ISIS): Winnipeg, MB, Canada, 2005.
21. Kamal, A.S.M.; Boulfiza, M. Durability of GFRP rebars in simulated concrete solutions under accelerated aging conditions. *ASCE J. Compos. Constr.* 2012, *15*, 473–481.
22. Benmokrane, B.; Wnag, P.; Pavate, T.; Robert, M. *Durability of FRP Composites for Civil Infrastructure Applications*; Whittles Publishing: Dunbeath, Scotland, 2006.]
23. American Society for Testing and Materials (ASTM). *Standard Test Method for Ignition Loss of Cured Reinforced Resins*; ASTM D 2584–13. ASTM: West Conshohocken, PA, USA, 2013.
24. Trejo, D.; Gardoni, P.; Kim, J.J. Long-term performance of GFRP reinforcement embedded in concrete. *ACI Mater. J.* 2011, *108*, 605–613.
25. Gardoni, P.; Trejo, D.; Kim, Y.H. Time-variant capacity model for GFRP bars embedded in concrete. *ASCE J. Eng. Mech.* 2013, *139*, 1435–1445.
26. Kim, Y.H.; Trejo, D.; Gardoni, P. Time-variant capacity and reliability of GFRP-reinforced bridge decks. *ASCE J. Compos. Constr.* 2012, *16*, 359–370.
27. American Concrete Institute (ACI). *Building Code Requirements for Structural Concrete and Commentary*; ACI 318, 318–11. ACI: Farmington Hills, MI, USA, 2011.
28. Park, Y. Long-Term Performance of GFRP Reinforced Concrete Beams and Bars Subjected to Aggressive Environments. *Ph.D. Thesis*, The University of Texas at Arlington, Arlington, TX, USA, May 2012.
29. Toutanji, H.A.; Saafi, M. Flexural behavior of concrete beams reinforced with glass fiber-reinforced polymer (GFRP) bars. *ACI Struct. J.* 2000, *97*, 712–719.
30. Bischoff, P.; Gross, S. Equivalent moment of inertia based on integration of curvature. *ASCE J. Compos. Constr.* 2011, *15*, 263–273.

31. Grace, N.; Soliman, A.; Abdel-Sayed, G.; Saleh, K. Behavior and ductility of simple and continuous frp reinforced beams. *ASCE J. Compos. Constr.* 1998, *2*, 186–194.
32. Jaeger, L.G.; Mufti, A.A.; Tadros, G. The concept of the overall performance factor in rectangular-section reinforced concrete members. In Proceedings of the 3rd International Symposium on Non-Metallic (FRP) Reinforcement for Concrete Structures, Sapporo, Japan, 14–16 October 1997; pp. 551–559.
33. Naaman, A.; Jeong, S.M. Structural ductility of concrete beams prestressed with FRP tendons. In non-metric (FRP) reinforcement for concrete structures. In Proceedings of the 2nd International RILEM Symposium (FRPRCS-2), Ghent, Belgium, 23–25 August 1995; pp. 379–386.
34. Zou, P. Flexural behavior and deformability of fiber reinforced polymer prestressed concrete beams. *ASCE J. Compos. Constr.* 2003, *7*, 275–284.
35. Abdelrahman, A.; Tadros, G.; Rizkalla, S. Test model for the first canadian smart highway bridge. *ACI Struct. J.* 1995, *92*, 1–9.
36. Newhook, J.; Svecova, D. *Reinforcing Concrete Structures with Fibre Reinforced Polymers*; The Canadian Network of Centres of Excellence, ISIS Canada, University of Manitoba: Winnipeg, Canada, 2007.

Chapter 2

Active In-Database Processing to Support Ambient Assisted Living Systems

Wagner O. de Morais, Jens Lundström, and Nicholas Wickström

School of Information Science, Computer and Electrical Engineering, Halmstad University, Box 823, Halmstad 30118, Sweden.

ABSTRACT

As an alternative to the existing software architectures that underpin the development of smart homes and ambient assisted living (AAL) systems, this work presents a database-centric architecture that takes advantage of active databases and in-database processing. Current platforms supporting AAL systems use database management systems (DBMSs) exclusively for data storage. Active databases employ database triggers to detect and react to events taking place inside or outside of the database. DBMSs can be extended with stored procedures and functions that enable in-database processing. This means that the data processing is integrated and performed within the DBMS. The feasibility and flexibility of the proposed approach were demonstrated with the implementation of three distinct AAL services. The active database was used to detect bed-exits and to discover common room transitions and deviations during the night. In-database machine learning methods were used to model early night behaviors. Consequently, active in-database processing avoids transferring sensitive data outside the database, and this improves performance, security and privacy. Furthermore, centralizing the computation into the DBMS facilitates code reuse, adaptation and maintenance. These are important system properties that take into account the evolving heterogeneity of users, their needs and the devices that are characteristic of smart homes and AAL systems. Therefore, DBMSs can provide capabilities to address requirements for scalability, security, privacy, dependability and personalization in applications of smart environments in healthcare.

INTRODUCTION

Storage is an important and required functionality in continuous, long-term, home-based monitoring systems, and the database management system (DBMS) is the most common, but not fully exploited, component among software architectures underpinning smart environments, such as smart homes, and ambient assisted living (AAL) systems.

As an extension [1] and an alternative to existing platforms supporting the development of smart homes and AAL systems, this work proposes a database-centric architecture that explores the capabilities of DBMSs beyond those of data management.

Background

Home care has been suggested to be a sustainable alternative to traditional care, because it has the potential to prevent unnecessary acute or long-term institutionalization and to enable individuals to stay in their homes and communities for as long as possible [2]. Similar to other countries in Europe, in Sweden, county councils and municipalities provide medical, social and personal care services for care beneficiaries in their own homes. Most people receiving home care services are old or disabled individuals living alone.

Home care visits are normally planned, but when social alarm devices are offered to care beneficiaries, unplanned emergency response visits also occur. A social alarm is a portable device and includes a push-button to alert a care unit. Social alarms can include also a movement sensor that automatically triggers an alarm upon inactivity. The device is commonly worn as a wrist-watch or as a pendant necklace.

Although most home care services are provided during the daytime, some individuals require assistance during the night. In a nighttime visit, a night patrol team can help with medication, diapering, toileting and repositioning in bed [3]. However, supervision visits are more common, *i.e.*, the night patrol team, without waking up the resident, checks if the person is in bed, breathing and doing fine. In Halmstad, Sweden, about 200 homes are visited and around 780 km are driven each night by caregivers providing nighttime home care [4].

Even though home care services prevent the institutionalization of many individuals [3], there are a number of issues that will likely limit their efficiency and effectiveness in the near future. By the year 2050, about 27% of the European population is expected to be of the age 65 years and above, and in Sweden, older adults will account for 23% of the Swedish population [5]. While many individuals will remain healthy and independent into late adulthood, others will be highly dependent on informal or professional care [6]. Consequently, the demand for home care services will drastically increase, and as it currently stands, the healthcare system is not prepared to address these demands, mostly due to the shortage of professionals specializing in geriatric care [7] and nighttime caregiving [3,8].

When it comes to nighttime home care, not all individuals receiving such services are actually in need of it, because they are still relatively independent and can use the social alarm device to request assistance if it is ever needed [4]. Furthermore, some care beneficiaries report being awakened by the night patrol supervision visit, and these individuals often trigger their alarm within minutes after a supervision visit [4].

Smart home technologies can enhance or complement home health care and have been shown to be integral parts of a cost-effective healthcare system [9, 10]. A smart home provides a home-based infrastructure that integrates network-enabled devices with different capabilities to offer advanced functionalities to the residents. Traditionally, smart homes have included solutions to enhance the comfort and safety of residents, as well as systems to manage and conserve energy [11, 12]. However, over the past several years, there has been an increased interest in using the pervasive infrastructure of smart homes to support aging in place and AAL.

Systems targeting aging in place and AAL aim to support older or disabled individuals with services that: (1) promote healthier lifestyle and enhanced quality of life; (2) enable early disease detection and treatment compliance; (3) support informal and professional caregiving; and (4) enable individuals to live independently for a longer time in their own homes [13, 14].

The current practice of assessing the nature of chronic diseases is limited to clinic-based assessments scheduled at discrete points in time, and the management of illnesses is limited to a few medical visits and to self-reports [15]. The collection and analysis of functional, safety, security and physiological parameters, as well as cognitive and social support are the most common smart home applications in healthcare [16].

In-home health monitoring provides accurate and reliable long-term data to support better decision making, better understanding of aging and illnesses, the prevention and management of chronic diseases, healthier attitudes and behaviors and the conservation of healthcare resources [15,17]. Moreover, the long-term storage of health-related information enables the use of data mining methods that can reveal unknown patterns or relationships that can indicate the onset of a health-related problem [18].

Smart homes and AAL systems are complex to build, use and maintain [19]. One factor contributing to such complexity is the inherent diversity that is characteristic of smart homes and that leads to technical issues related to personalization, integration, interoperation, extensibility and dependability. Individuals have needs, preferences, habits and adverse health conditions that differ and evolve over time. Home environments also differ, and heterogeneous technologies, such as sensors and actuators, are employed in these systems. These distinct devices are provided by different manufacturers, and they operate and communicate with different standards and protocols. Thus, there is no universal arrangement of devices to fit every home environment.

The acceptance of smart homes and AAL systems is also an issue and is directly linked with the system's ability to address an individual's evolving needs, as well as their concerns for privacy, security and dependability [20]. Regarding privacy, not all individuals will accept technologies that monitor all aspects of their lives. Cameras, for example, are perceived as invasive technologies. Moreover, collected data from such systems are very sensitive. In the same way that data analysis of stored data can predict the onset of a health-related problem, data analysis could also predict the predisposition of a person to commit a crime [21]. As a consequence, there are different issues related to data security, such as who is going to use or have access to the data and how and where the data is going to be processed, stored and used. Concerns related to dependability are associated with trust, e.g., can users rely on the system and what if the system stops working altogether.

Related Work

A number of smart homes and AAL projects have been developed over the past several years (reviewed in [22, 23]) along with the technical infrastructures that serve as foundations for AAL applications [24]. Although some architectural aspects are common among existing smart environments and AAL platforms, there is still no widely adopted method for developing these systems [25]. Different software architectures have been proposed for the smart environment and AAL domains, including service-oriented architecture (SOA), service-oriented device architecture (SODA), peer-to-peer architecture (P2P), event-driven architecture (EDA), component and connector (C2), multi-agent system (M.A.S) and blackboard. However, as discussed in [26], none of them can perfectly fit the requirements for AAL systems, specifically the requirement for integration [26].

Sensors and actuators provide the means for perceiving and controlling the environment. These devices, among others, are provided by different manufacturers and operate and communicate through different standards and protocols. The open-service gateway initiative (OSGi) framework is commonly used to abstract and integrate devices, such as sensors and actuators, as well as to create service-oriented applications.

Several projects have adopted platforms or middlewares based on SOA and built on top of the OSGi service framework. The Gator Tech Smart House [27], PERSONA (PERceptive Spaces prOmoting iNdependent Aging) [28], SOPRANO (Service Oriented PRogrammable smArt enviroNments for Older Europeans) [29] and universAAL (UNIVERsal open platform and reference Specification for Ambient Assisted Living) [30] are examples of smart homes and AAL projects based on SOA and OSGi.

Current infrastructures supporting smart environments and AAL solutions typically implement the domain logic along with methods for data analysis, data mining and machine learning, as well as the mechanisms for security and privacy at the application, service or middleware layers (Figure 1a).

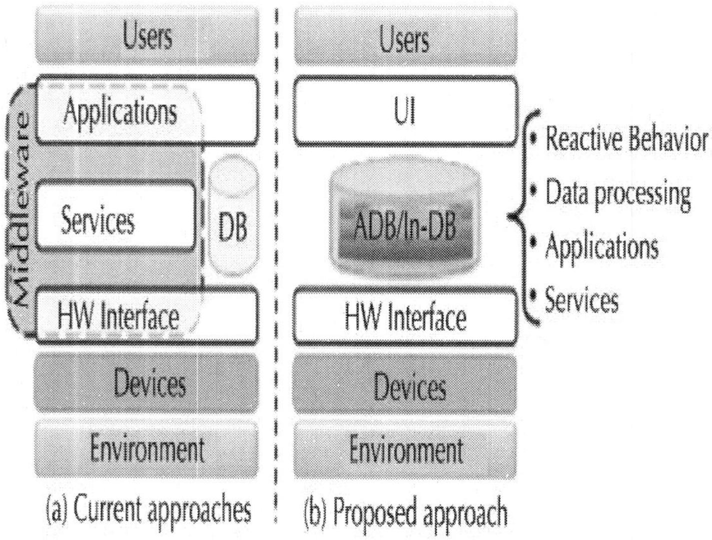

Figure 1: (**a**) Existing infrastructures supporting smart environments and AAL systems perform data processing at different layers; (**b**) in the proposed database-centric architecture, the reactive behavior and data processing are integrated and performed within the database management system (DBMS). Notation: ADB, active database; DB, database; In-DB, in-database processing, HW, hardware; UI, user interface.

Modern DBMSs—such as PostgreSQL [31]—provide mechanisms that can be utilized to address important requirements for data processing and analysis, security, privacy, dependability, extensibility and scalability in smart home and AAL systems. Such mechanisms have not been explored by current smart environments and AAL infrastructures that employ DBMSs exclusively for data storage and retrieval.

Approach and Contribution

In response to the challenges previously described and as an alternative to current approaches, this work presents a database-centric system architecture that exploits mechanisms provided by DBMSs to support the development of AAL applications. The aim is to push the reactive behavior and the data processing, which are commonly implemented at different software layers within existing architectures, into the DBMS (Figure 1b).

This work exploits active databases to detect and respond to events taking place in the home environment, such as bed-exits. The extensibility capabilities of DBMSs, which are mostly provided by user-defined functions, are also explored in this work to perform in-database processing. This means that the domain logic (e.g., for detecting and responding to emergencies) is integrated into the DBMS itself. Three distinct AAL services—bed-exit detection, discovery of common room transitions and behavior modeling—are implemented using the proposed database-centric architecture and are evaluated with a dataset collected in real homes from older individuals living alone.

Active databases and in-database processing avoid transferring sensitive data outside the database. Moreover, the domain logic is centralized into the DBMS and managed on the fly.

The remainder of this paper is organized as follows. An overview of the capabilities of DBMSs, other than data management, is presented in Section 2. Section 3 describes a motivating scenario for AAL applications. The proposed database-centric architecture and its main components are presented in Section 4 and are evaluated with the development of three home-based healthcare services in Section 5. Conclusions are presented in Section 6.

CAPABILITIES OF DATABASE MANAGEMENT SYSTEMS

Traditionally, DBMSs are passive components in architectures supporting smart environments and AAL solutions and are employed exclusively to store and manage data for later retrieval. The SQL (Structured Query Language) language is used solely for specifying the database schema and for accessing or manipulating data. However, DBMSs can do much more than data management.

DBMSs incorporate active rule processing mechanisms in the form of database triggers. These provide an event-driven architecture that enables the DBMS to monitor and react to events taking place inside or outside of the database, for example, to enforce referential integrity or to react to sensor data being inserted into the database, respectively.

Moreover, DBMSs enable developers to implement new procedures, functions and data types that are stored within the DBMS. DBMSs also promote mechanisms for controlling security and privacy. DBMSs are very dependable systems, mostly due to high-availability, robustness and reliability, and they enable changes in the domain logic, reactive behavior and security policies to be managed on the fly. This facilitates the system's scalability, maintainability and personalization, because changes in software applications connected to the DBMS are not required [32].

Although the aforementioned capabilities are present in the most widely-used commercial (e.g., Oracle, Microsoft SQL Server and IBM DB2) and open-source (e.g., PostgreSQL and MySQL) DBMSs, the database-centric architecture presented in this work focuses only on the capabilities provided by PostgreSQL [31].

Active Databases

The SQL language enables the creation of database triggers that provide an in-database event-driven architecture to detect and respond to events, such as data manipulation operations, such as table insertions and updates. Database triggers are event-condition-action (ECA) structures—commonly referred to as active rules—meaning that when an event occurs, the condition is evaluated, and if it holds, an action is executed. The action can be executed before or after a data manipulation operation, for example, after a table insertion and/or update.

DBMSs exploiting active rules are called active databases [33]. An active database can monitor and respond to specific circumstances of relevance to an application in a timely manner [33]. For example, active rules can react to incoming sensor data to control smart environments [34]. Active databases can also prevent client applications from periodically querying (polling) the database for data changes. Periodic polling mechanisms can be inefficient (too many queries due to a short polling interval) and inaccurate (delayed response due to a long polling interval). To notify client applications about the occurrence of a certain event, such as a data change, active database systems can make use of external or built-in inter-process communication mechanisms. Such an approach requires the client application to be always connected to the DBMS and to subscribe to notifications published by the DBMS.

PostgreSQL, for example, provides a built-in asynchronous publish-subscribe mechanism for inter-process communication using the NOTIFY (publish), LISTEN (subscribe) and UNLISTEN (unsubscribe) commands.

SQL Extensions

DBMSs enable the SQL language to be extended with user-defined types (UDTs), user-defined aggregates (UDAs), user-defined functions (UDFs) and stored procedures (SPs). UDTs, UDAs, UDFs and SPs can subsequently be included in SQL statements and queries. Moreover, the actions invoked by database triggers are commonly implemented as UDFs or SPs. UDFs and SPs enable in-database processing and analytics—*i.e.*, the semantics of applications, statistical models and machine learning techniques—to be integrated and performed within the DBMS. SQL extensions, including database triggers, are implemented in SQL language or using database vendor-specific procedural languages, such as PL/pgSQL (procedural language for PostgreSQL), Python variants [31] and C language.

PostGIS [35], for example, is a free and open source database extension that adds spatial and geographic objects for PostgreSQL. Advanced algorithms, such as methods for statistical analysis and machine learning, can also be integrated into modern DBMSs. For example, MADlib [36] is an open-source library that adds in-database analytical capabilities for PostgreSQL. The MADlib library supports established methods for supervised learning (linear and logistic regression, decision trees and support vector machines), unsupervised learning (k-means clustering and association rules) and descriptive statistics, and it comes with support modules that provide array operators and probability functions among many other methods [36].

Database extensions are stored into the DBMS and are managed on the fly without requiring system restarts. In-database processing facilitates code reuse and maintainability, avoids data movement and improves performance and security. Performing data processing inside the DBMS is more efficient than with external data mining programs [37,38]. The in-database implementation of different statistical models and machine learning techniques, along with their advantages, can be found and are discussed in [37–42].

Security and Privacy

In addition to active databases and in-database processing, which avoid transferring sensitive data from the database to external applications, DBMSs provide other mechanisms to enforce data security and privacy, such as authentication and authorization.

DBMSs support authentication mechanisms that are used to check and confirm the identity of a user, device or software application trying to access database resources. Besides password-based authentication, DBMSs, such as PostgreSQL, enable authentication methods and protocols, such as Lightweight Directory Access Protocol (LDAP) authentication, the Kerberos network authentication protocol, and Secure Sockets Layer (SSL) certificates, among others.

DBMs also support authorization mechanisms that are used to manage and control users' access permissions to database resources. PostgreSQL manages database access permissions using the concept of roles that can be attributed to a DBMS user or to a group of DBMS users [31].

MOTIVATING SCENARIO: THE "TRYGG OM NATTEN" (SAFE AT NIGHT) PROJECT

The "Trygg om natten" (Safe at night in Swedish) project was conducted in Halmstad, Sweden and explored how technology could assist care beneficiaries and caregivers during nighttime supervisions [4]. The study also focused on how technology was perceived by the participants in terms of integrity and acceptance.

The criteria for selecting participants for the project were that individuals had to be beneficiaries of nighttime supervisions, live alone in their own house or apartment without pets and sustain some level of independence, such as for showering, dressing, eating, functional mobility and personal and toilet hygiene. In addition, an approval of the night patrol team was also required. Individuals diagnosed with some type of dementia or not able to give informed consent were excluded.

In total, 15 out of 30 nighttime supervisions beneficiaries (2 men and 13 women) with an average age of 82 years participated in the project. Ten participants lived in apartments and five in houses.

The home of each participant was equipped with five types of sensors (Table 1) that were active from 10 p.m. until 6 a.m. every night for approximately 14 days.

Table 1: Sensors used in the "Trygg om natten" (Safe at night) project [4]

Type	Purpose	Quantity	Output
Passive infrared (PIR)	Capture human motion	3–5	Binary
Quasi-electric film (Emfit)	Capture bed exits	1	Binary
Magnetic	Capture door openings	1	Binary
Inertial sensor	Capture human activity (wearable)	1	Binary
Load cell	Reference for the Emfit sensor	1	24-bit value

Two data collections were discontinued during the project, one due to the illness of the participant and another because the participant no longer had the need for nighttime supervision.

Figure 2 illustrates possible placements of different types of sensors within the home environment. The Emfit Bed Sensor was used in the project as the main method to detect bed exits. One strain-gauge load cell was placed at the top-left corner support of the participant's bed to serve as a reference for the Emfit Bed Sensor. Bed entrances and exits, as well as presence in bed, were derived from the measured weight data. Motion sensors in different locations in the home captured human movement in the bedroom, living room, bathroom and kitchen. A magnet sensor installed in the front door monitored whether the front door was opened or closed. The intent with the magnet sensor in the front door was to capture nighttime supervision visits. Except for the load cell, all of the other sensors transmitted the measured data wirelessly to a low-power, fanless, miniature host computer located under the bed.

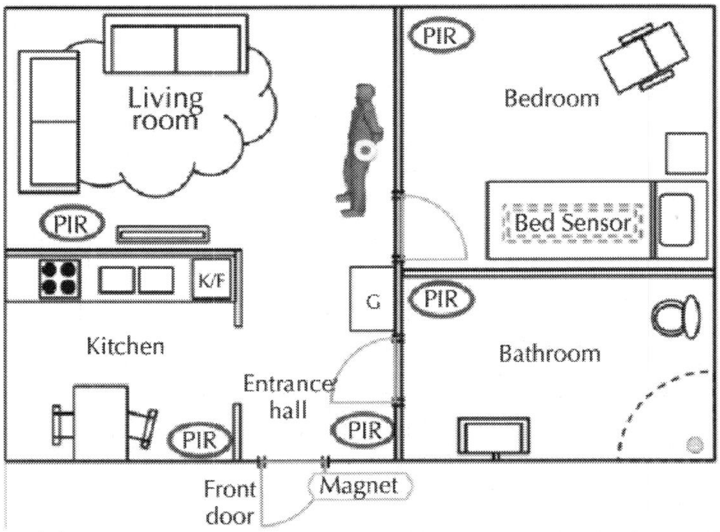

Figure 2: Example of a sensor setup for a given home environment. PIR denotes passive infrared motion sensors; the magnet to capture door openings; the bed sensor to detect bed exits; the resident wears a social alarm. A load cell to measure weight is placed on the top-left leg support of the bed.

The load cell was connected to an analog-to-digital converter that was connected to a USB port of the host computer. The study was granted ethical approval from the central ethical review board. One of the outcomes of the study was a set of requirements and specifications for AAL services, particularly those related to nighttime caregiving. The dataset collected during the "Trygg om natten" project was used in this work to evaluate the proposed services.

DATABASE-CENTRIC ARCHITECTURE TO SUPPORT AMBIENT ASSISTED LIVING SYSTEMS

This section presents how different DBMSs capabilities fit together in the proposed database-centric system architecture to support smart homes and AAL systems.

Figure 3 summarizes the framework in which the proposed system operates and its main components, described in the next subsections.

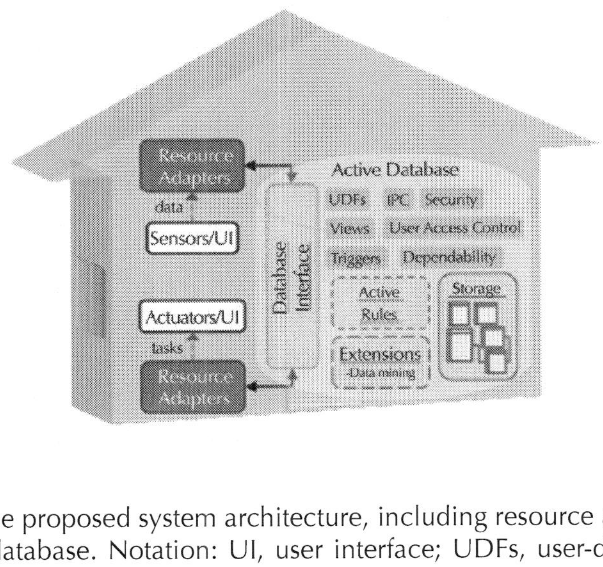

Figure 3: The proposed system architecture, including resource adapters and the active database. Notation: UI, user interface; UDFs, user-defined functions; IPC, inter-process communication.

Resource Adapters

As there is still no adopted standard for communicating with and integrating devices and applications inside smart homes [43], resource adapters have been designed to abstract heterogeneous hardware technologies (sensors and actuators) and software technologies (user interfaces and files) in order to facilitate their integration and interoperation into the system. Resource adapters resemble context widgets and context services [44], but with fewer responsibilities (no data aggregation or peer-to-peer communication). Resource adapters encapsulate the underlying implementation of different communication protocols and abstract resource-specific data formats. Recovery from faults, such as communication disconnections, can also be provided. Resource adapters serve as a gateway between the environment and the DBMS and are implementable in different programming languages, such as C# and Python. Resource adapters stream data acquired by sensors or entered by the user into the database. They also control actuators and user interfaces in response to commands received from the database. Resource adapters communicate with the database through the database interface (Figure 3), and the DBMS employs inter-process communication mechanisms to communicate with resource adapters. Therefore, resource adapters keep an open connection with the DBMS and subscribe to specific event channels.

Active Database

The active database (Figure 3) includes several modules that are used as follows.

Storage

The storage module includes the tables for storing sensor data, processed information and meta-data (location, capabilities and configuration) of the hardware and software resources that are present in the environment. Developers implementing resource adapters do not have access to the internal storage model. They are provided instead with a database interface.

Database Interface

The internal database model is protected from direct access by the database interface module that exposes data access (selections) and manipulation (insertions, updates and deletions) using views and UDFs. Listing 1 shows an example of such an approach.

Listing 1: UDF written in PL/pgSQL for inserting converted weight samples into table *weight*

```
1.  CREATE FUNCTION weight_insert ( adc_out integer , ts timestamp )
2.  RETURNS boolean AS $$
3.  DECLARE
4.  voltage_weight_ratio numeric := −41943.0;
5.  weight_sample numeric ;
6.  BEGIN
7.  weight_sample := adc_out / voltage_weight_ratio ;
8.  INSERT INTO weight VALUES ( weight_sample , ts );
9.  RETURN true ;
10. END;
11. $$ LANGUAGE PLPGSQL
```

The UDF named *weight_insert* abstracts the insertion into table *weight* and is also used to process the input parameters. The UDF *weight_insert* receives two parameters, the output of the analog-to-digital converter (*adc_out*) and timestamp (*ts*). In Listing 1, Line 7, the voltage-to-weight ratio (*voltage_weight_ratio*) variable is used to convert the readout value (*adc_out*) to weight (*weight_sample*), which is later inserted into table *weight*. Such an approach facilitates changes in the logic, such as in the voltage to weight conversion, because it is performed on the fly and does not require modifications or recompilations of resource adapters.

To notify resource adapters about data changes or events, the active database makes use of built-in mechanisms in PostgreSQL for inter-process communication (NOTIFY and LISTEN commands), and this prevents resource adapters from periodically querying (polling) the database.

Active Rules

The reactive behavior in the system is supported by the active rules module. In conjunction with sensors and actuators, active rules implemented as database triggers (Listing 2) can monitor and react to events happening in the environment.

Listing 2: A database trigger monitors when a sequential sample identifier (*sample_id*) of a first in, first out (FIFO)-type of table (*weight_fifo*) wraps around to execute an action (*check_presence_absence*)

```
1.  CREATE TRIGGER weight_fifo_after_insert
2.    AFTER INSERT
3.    ON weight_fifo
4.    FOR EACH ROW
5.    WHEN ( NEW. sample_id == 40 )
6.    EXECUTE PROCEDURE check_presence_absence ( );
```

In Listing 2, the ECA rule represented by the trigger *weight_fifo_after_insert* is associated with the table *weight_fifo* and fires after table insertion events. If the condition specified by the Boolean expression in Listing 2 Line 5 is satisfied, the action—*check_presence_absence*—is executed. Because the analog-to-digital samples the load cell at 80 Hz, the trigger fires every half second or every 40th insertion and is intended to detect bed entrances and exits.

Database Extensions

Active rules invoke actions that can be functions added by database extensions, such as MADlib [36], or can be user defined. These functions implement both short-term and long-term types of services. Short-term services are those that respond to simple events, such as generating an alarm indicating a bed exit. Long-term services are defined as services requiring datasets collected over a longer period of time and the analysis of patterns in such data, for example, to gain knowledge about preferences or to detect abnormal behaviors [45]. Because sensitive data are involved in the data processing, implementing the methods for such analysis into the DBMS itself avoids data movement and leads to better performance and security.

Security

Table 2 presents possible access privileges according to different roles in the system (similarly to [46]). The owner can grant or revoke the access privileges of other system users. Software developers creating resource adapters are granted execute permission on specific UDFs within the database interface.

Table 2: Access privileges according to different roles

Role	Access Level			
	View	**Add**	**Modify**	**Administer**
Owner	X	X	X	X
Family	X	X	X	X
Healthcare	X	X	X	
Other users				
Devices		X		

EXPERIMENTAL RESULTS: IN-DATABASE SERVICES SUPPORTING AAL SYSTEMS

Three distinct AAL services for home-based health monitoring, inspired by the "Trygg om natten" project (Section 3), are presented and implemented following the proposed architecture.

To accommodate the proposed architecture, a database server was configured in a computer running CentOS 6.4 with PostgreSQL (version 9.2.3) and the MADlib [36] library extension. To implement the proposed services, additional tables were created to store temporary and derived data, such as descriptive statistics and transition matrices. A separate computer running MS Windows 7 hosted resource adapters (implemented in C#) that read the measurements from the "Trygg om natten" dataset files to the corresponding database. The dataset from a single care beneficiary, who was an active man, living alone in his own apartment and receiving daytime home care services and nighttime supervision, was selected to present the implementation results.

Detection of Bed Presence and Absence

A service to detect presence in bed can enable the night patrol team to remotely check if individuals are in bed, so as not to disturb their sleep. Voluntary and involuntary body movements create disturbances in the load cell signal that are not present when the bed is unoccupied or is loaded with static weight. Figure 4 presents the standard deviation of a weight signal measured by a load-cell sensor in the moments before the person left the bed. By analyzing the measured weight and its standard deviation, a method to detect the presence or absence of a person in bed can be implemented as an active rule (trigger) that monitors the table in which the measured weight is stored.

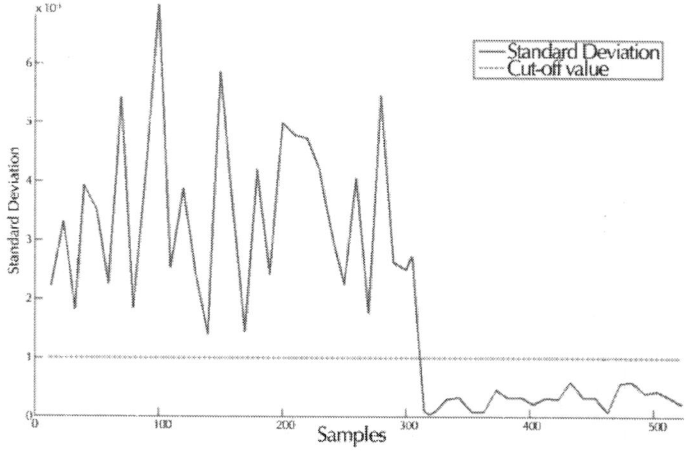

Figure 4: A cut-off value can separate the standard deviation of the measured weight signal into in-bed signals and out-of-bed signals.

The active rule triggers every half second and invokes a UDF that checks for bed exits and entrances. The condition (Equation (1)) for the detection consists of checking intervals in which the median (m_w) and the standard deviation ($_w$) of the weight signal are greater than the respective estimated thresholds for the mean value (O_{mw}) and standard deviation (O_w).

$$\text{Presence} = \left(\left(\sigma_w \geq O_{\sigma_w} \right) \text{AND} \left(m_w \geq O_{m_w} \right) \right) \quad (1)$$

The mean value and standard deviation of the weight signal are calculated with a moving window with the last 40 inserted weight samples (approximately half a second or half of the signal sampling rate, which was 80 Hz). Smaller window sizes can lead to high granularity that makes it difficult to find the separating threshold, and larger window sizes can delay the detection of bed entrances and exits.

A method for finding a threshold in a signal (*i.e.*, binarizing) is the Otsu algorithm [47], which maximizes the between-cluster distance when dividing the distribution of values into two clusters, for example, the presence and absence clusters. For each individual, corresponding thresholds have been calculated.

For the selected individual, 27 bed presences and 16 bed absences were detected by the active rule based on measured weight. To identify true and false positives, the dataset containing load cell signals was manually labeled and served as a baseline for comparison.

All bed presence and absence detections were validated as true positives. Bed absence detections outnumbered bed presence detections, because on many occasions, the individual left the bed after the sensors became inactive at 6 a.m.

The proposed approach to detect bed exits and entrances also detected more bed-exit events than the bed-exit detection provided by the Emfit Bed Sensor. Figure 5 presents one missed and one nonexistent bed exit using the Emfit Bed Sensor. For the same individual, the Emfit Bed Sensor missed approximately 60% of all bed exits. Such a mismatch might be caused by the antidecubitus mattress that the individual was using to prevent and treat pressure sores.

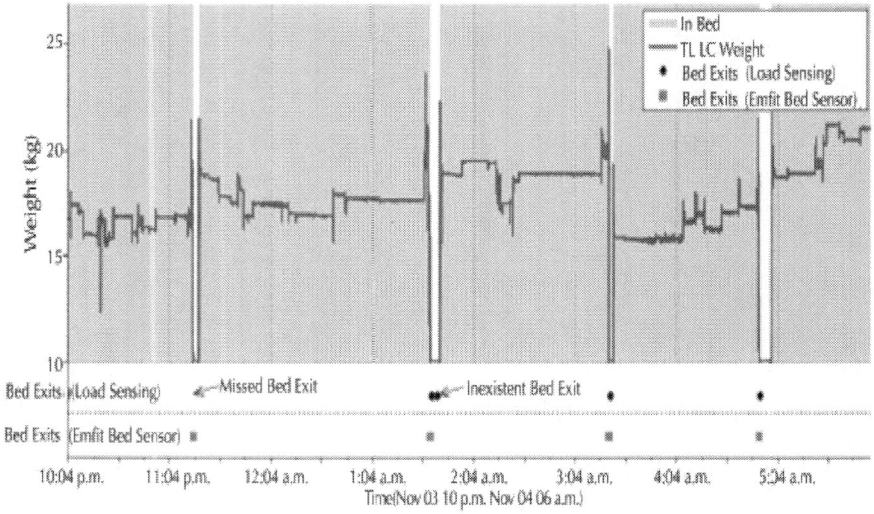

Figure 5: Bed entrances and exits are accurately detected by the active rule, while the bed-exit detection provided by the Emfit Bed Sensor misses bed exits or generates nonexistent bed exits. Because the sensors were active from 10 p.m. until 6 a.m., it was not possible to detect when the individual went to bed or when he left. TL LC denotes top-left load cell.

The overall approach avoids raw load cell data, which exposes private and sensitive information, from being transferred and processed outside of the DBMS. In this system, several resource adapters can subscribe to the service and are notified when bed entrances and exits are detected.

Common Event Transitions during the Night

The purpose of this service is to enable the detection of anomalies by discovering simple associations between presence detections in the bathroom, living room, kitchen, entrance hall and bed. Strong associations indicate common room transitions and room activity, and deviations from such associations can enable the detection of anomalies.

A method for finding such expected patterns in sequences of events (*i.e.*, sequential data mining [48]) is by estimating the probability $p(e_y|e_x)$ of one event e_x being followed by another type of event e_y

(similar to [49]). By considering only the previous detected event, a transition matrix can be computed online for each individual using an active rule. Each element in the transition matrix *P* contains the probability of event e_i being followed by event e_j, and this is denoted as $P_{ij}(e_j|e_i)$, which is also referred to as the confidence in association rules [50]. The transition matrix can be visualized as a graph by plotting associations over a certain confidence threshold.

An active rule monitors incoming events from all sensors (Table 1 in Section 3) and updates the transition matrix table, which describes the transition probability of events happening during the night. The computation of statistics, such as the mean and standard deviation of the transition time between two events, is also triggered by the rule. Bed-exit events generated by the previously proposed active rule were used due to the higher accuracy than bed-exit events detected by the Emfit Bed Sensor (Figure 5 in Section 5.1).

Figure 6 presents likely transitions of events in the home environment of the selected subject. An observation from the figure is that when the observed individual leaves the bed, the most likely event is a visit to the bathroom. Such a transition takes an average of 7 min with a standard deviation of 7 min.

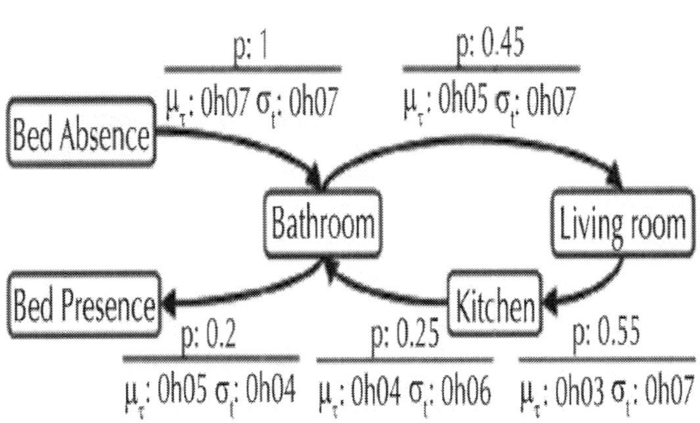

Figure 6: Transition probabilities (*p*) of events for a confidence threshold of 0.2. Mean (μ_t) and standard deviation ($_t$) of the transition time (normally distributed).

The knowledge provided by the transition matrix can be used to detect anomalies during future nights. Anomaly detection mechanisms can also be implemented with active rules. Because health-related conditions evolve over time and because health changes might not be evident in the short-term, the amount of stored data to be processed increases by a large amount every day. Therefore, in-database sequential data mining avoids transferring stored long-term data to external data analysis tools to update transition probabilities.

Modeling of Early Night Behavior Using Decision Trees

Another way to model transitions is with a service that models typical sensor triggering transitions over a certain time span during the night. Such a service could help to discover changing trends in the level of independence of the individual being monitored.

For this service, a decision tree using the C4.5 implementation in MADlib was trained with data from a single individual to discriminate between the time period from 10 p.m. to midnight (TPI denotes Time Period I) and the period from midnight to 6 a.m. (!TPI denotes not Time Period I). The training data consisted of 15 features that were computed for each observation by processing a sliding window with a width of 20 min over the 14 days of collected data. No feature selection has been applied due to the rather low number of features used. This process resulted in training data with approximately 300 observations.

The events in the collected data are denoted as bathroom (Ba), kitchen (K), hallway (H), and living room (L), and each event represents activity in a certain room. Other events include inactivity registered from the wearable inertial sensor (I), door openings (D) and bed entrances and exits (Bin and Bout, respectively), which are computed using the proposed active rule for detecting bed entrances and exits. The features used in the calculations are the type of sensors that fired in the last four events and are denoted as *event at time t*. The transition time between the four last events for the window is computed as *Et*(*t*, *t* – *k*), where *k* is the number of previous events. The number of each type of event and the lack of events (denoted by N) in a window are also computed.

The generated decision tree for the same individual is shown in Figure 7. Thick edges represent where the majority of data points were concentrated. The tree hierarchy reflects variable importance. Figure 7 illustrates (by hierarchy and bold edges) that the last occurring event is important. An example of this is the lack of events (N) in a window, and the presence or absence in bed were the most informative, while discriminating between TPI and !TPI.

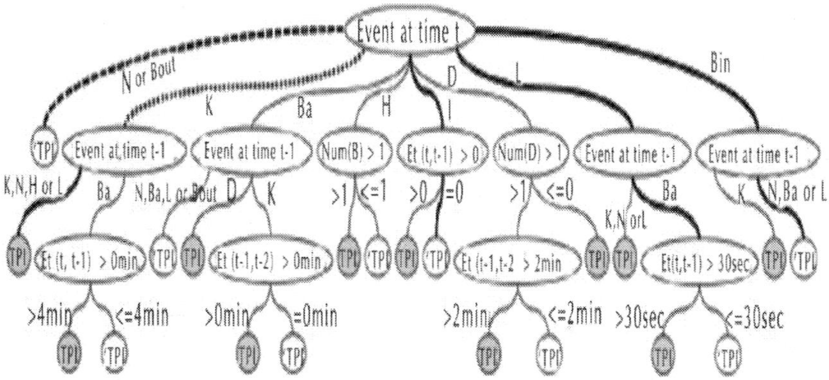

Figure 7: A decision tree distinguishes different time periods during the night. Notation: Ba, bathroom; K, kitchen; H, hallway; L, living room; I, inactivity; D, door openings; Bin, bed entrances; Bout, bed exit; N, lack of events. TPI, Time Period I; !TPI, not Time Period I.

One interpretation of the model is illustrated by the dashed edge from the root node. This link revealed that the individual was more likely to be active during the modeled time period TPI than the rest of the night (!TPI). Moreover, the dotted edge shows that the individual was active in the kitchen, hallway and living room during TPI. In order to validate the decision tree model, a 10-fold cross-validation was performed, and a mean accuracy of 81% was achieved. The accuracy shows that, despite the complexity of human behavior, the model is able to explain key features of the early night that could be used when analyzing deviations in long-term trends.

Similar to updating transition probabilities, in-database retraining of decision trees avoids data movement, and this promotes privacy. The processing time to create or update a probability matrix and to train a decision tree can be negligible for a small dataset, but quite significant for a dataset containing months of stored data.

CONCLUSIONS

This work has shown how different capabilities of DBMSs (e.g., triggers, user-defined functions and existing database extensions for in-database analytics) fit together in a database-centric architecture intended to support the development of home-based healthcare applications. The proposed software architecture represents an alternative to existing platforms supporting the development of smart homes and AAL systems.

DBMSs are mature and dependable technologies and provide mechanisms that can address the processing, security, privacy and personalization requirements of smart homes and AAL systems. These mechanisms, however, are not fully exploited in current smart home and AAL infrastructures.

In the system design presented here, database triggers are used to detect and respond to events taking place in the home environment. The event-driven architecture provided by active databases makes it possible to implement an in-database service to monitor an individual's presence or absence in bed, as well as to discover common room transitions and deviations during the night.

User-defined functions are exploited to perform in-database processing, *i.e.*, the domain logic is integrated into the DBMS itself. A database interface created with user-defined functions and views protects the internal database model against direct access. Existing DBMS extensions for data mining, such as MADlib, enable the development of services to model early night behaviors. Database roles can promote security by controlling user access to database resources, such as tables that contain private data.

The proposed and implemented AAL services, which have been validated with a dataset collected in real homes, reside within the database and avoid exporting sensitive data to external data analysis tools.

Therefore, active in-database processing avoids data movement from the DBMS to external applications. Such an approach can lead to improved performance, security and privacy while still benefiting from the on the fly management capabilities of DBMSs. Centralizing the domain logic into the DBMSs reduces code duplication, promotes code reuse and facilitates system maintenance and adaptability as the environment and individual needs evolve.

Although these are important system properties supported by the presented database-centric platform, the proposed approach requires developers to have knowledge of relational DBMSs, their features, such as for security, and procedural programming languages, such as PL/pgSQL and PL/Python [31], that extend the SQL standard.

Even though resource adapters keep an open connection with the DBMS to subscribe to notifications from the database, this is not a limitation, because resource adapters, for example running on battery powered devices, can connect and disconnect to the DBMS when necessary.

Up to now, besides the results reported in this work and in [1], the presented database-centric architecture has been used to develop a smart bedroom [34] and to integrate an autonomous mobile robot to such a smart environment [51]. Future work encompasses developing, deploying and evaluating smart environments that encompass a whole home environment and that provide actuation services to improve comfort, independence and medical care using the presented database-centric platform. To facilitate interoperation, semantic description for the environment, devices and user activities will also be investigated.

The authors would like to thank all of the participants in the "Trygg om natten" project [4], particularly Roland Thörner, Magnus Hållander and the care beneficiaries participating in the measurements. The authors would also like to thank Prof. Bertil Svensson and Prof. Antanas Verikas for their contributions to this work.

REFERENCES

1. De Morais, W.O.; Lundström, J.; Wickström, N. A Database-Centric Architecture for Home-Based Health Monitoring. In *Ambient Assisted Living and Active Aging*; Springer: Berlin, Germany, 2013; pp. 26–34.

2. Tarricone, R.; Tsouros, A.D. *Home Care in Europe: The Solid Facts*; WHO Regional Office: Copenhagen, Denmark, 2008.
3. Malmberg, B.; Ernsth, M.; Larsson, B.; Zarit, S.H. Angels of the night: Evening and night patrols for homebound elders in Sweden. *Gerontologist* 2003, *43*, 761–765.
4. Thörner, R.; Persson, M.; Eriksson, H.; Isaksson, A.; Lundström, J. Trygg om natten. In *Technical Report, Centre for Health Technology in Halland*; Halmstad University: Halmstad, Sweden, 2011. (In Swedish).
5. United Nations, Department of Economic and Social Affairs, Population Division. World Population Prospects: The 2012 Revision, Key Findings and Advance Tables (Working Paper No. ESA/P/WP.227). Available online: http://esa.un.org/wpp/index.htm (accessed on 16 December 2013).
6. Berleen, G.; Watson, G. *A Healthier Elderly Population in Sweden!*; Swedish National Institute of Public Health: Östersund, Sweden, 2004.
7. Kovner, C.T.; Mezey, M.; Harrington, C. Who cares for older adults? Workforce implications of an aging society. *Health Affairs* 2002, *21*, 78–89.
8. Arber, S.; Venn, S. Caregiving at night: Understanding the impact on carers. *J. Aging Stud.* 2011, *25*, 155–165.
9. Society, R. *Digital Healthcare: The Impact of Information and Communication Technologies on Health and Healthcare*; Technical Report. Royal Society: London, UK, 2006.
10. Kubitschke, L.; Cullen, K. *ICT & Aging-European Study on Users, Markets and Technologies*; Technical Report. Empirica and WRC on Behalf of the European Commission Directorate General for Information Society and Media: Bonn, Germany, 2010.
11. Mozer, M.C. The neural network house: An environment hat adapts to its inhabitants. Proceedings of the AAAI Spring Symposium Intelligent Environments, Menlo Park, CA, USA, 23–25 March 1998; pp. 110–114.
12. Das, S.K.; Cook, D.J.; Battacharya, A.; Heierman, E.O., III; Lin, T.Y. The role of prediction algorithms in the MavHome smart home architecture. *IEEE Wirel. Commun.* 2002, *9*, 77–84.

13. Dishman, E. Inventing wellness systems for aging in place. *Computer* 2004, *37*, 34–41.
14. Van den Broek, G.; Cavallo, F.; Wehrmann, C. *AALIANCE Ambient Assisted Living Roadmap*; IOS Press: Berlin, Germany, 2010; Volume 6.
15. Kang, H.G.; Mahoney, D.F.; Hoenig, H.; Hirth, V.A.; Bonato, P.; Hajjar, I.; Lipsitz, L.A. *In Situ* Monitoring of Health in Older Adults: Technologies and Issues. *J. Am. Geriatr. Soc.* 2010, *58*, 1579–1586.
16. Demiris, G.; Hensel, B.K. Technologies for an aging society: A systematic review of "smart home" applications. *Yearb. Med. Inform.* 2008, *3*, 33–40.
17. Paré, G.; Jaana, M.; Sicotte, C. Systematic review of home telemonitoring for chronic diseases: The evidence base. *J. Am. Med. Inform. Assoc.* 2007, *14*, 269–277.
18. Rashidi, P.; Cook, D.J. Mining and monitoring patterns of daily routines for assisted living in real world settings. Proceedings of the 1st ACM International Health Informatics Symposium, Arlington, VA, USA, 11–12 November 2010; pp. 336–345.
19. Eckl, R.; MacWilliams, A. Smart Home Challenges and Approaches to Solve Them: A Practical Industrial Perspective. In *Intelligent Interactive Assistance and Mobile Multimedia Computing*; Springer: Berlin, Germany, 2009; pp. 119–130.
20. Demins, G.; Thompson, H.J.; Reeder, B.; Wilamowska, K.; Zaslavsky, O. Using informatics to capture older adults' wellness. *Int. J. Med. Inform.* 2011, *82*, e232–e241.
21. The Economist. *New Rules for Big Data*; The Economist Newspaper Limited: London, UK, 2010. Available online: http://www.economist.com/node/15557487 (accessed on 10 December 2013).
22. Chan, M.; Estève, D.; Escriba, C.; Campo, E. A review of smart homes—Present state and future challenges. *Comput. Methods Programs Biomed.* 2008, *91*, 55–81.
23. Rashidi, P.; Mihailidis, A. A Survey on Ambient-Assisted Living Tools for Older Adults. *IEEE J. Biomed. Health Inform.* 2013, *17*, 579–590.

24. Fagerberg, G.; Kung, A.; Wichert, R.; Tazari, M.R.; Jean-Bart, B.; Bauer, G.; Zimmermann, G.; Furfari, F.; Potort, F.; Chessa, S.; *et al*. Platforms for AAL Applications. In *Smart Sensing and Context*; Springer: Berlin, Germany, 2010; pp. 177–201.
25. Augusto, J.C.; Callaghan, V.; Cook, D.; Kameas, A.; Satoh, I.; Saba, T.; Chorianopoulos, K.; Howard, N.; Cambria, E.; Gupta, V. Intelligent Environments: A manifesto. *Hum. Centr. Comput. Inf. Sci.* 2013, *3*, 1–18.
26. Becker, M. Assisted Living Systems-Models, Architectures Engineering Approaches. In *Software Architecture Trends and Promising Technology for Ambient Assisted Living Systems*; Karshmer, A.I., Nehmer, J., Raffler, H., Tröster, G., Eds.; Schloss Dagstuhl-Leibniz-Zentrum fuer Informatik, Germany: Dagstuhl, Germany, 2008.
27. Helal, S.; Mann, W.; El-Zabadani, H.; King, J.; Kaddoura, Y.; Jansen, E. The gator tech smart house: A programmable pervasive space. *Computer* 2005, *38*, 50–60.
28. Fides-Valero, Á.; Freddi, M.; Furfari, F.; Tazari, M.R. The PERSONA Framework for Supporting Context-Awareness in Open Distributed Systems. In *Ambient Intelligence*; Springer: Berlin, Germany, 2008; pp. 91–108.
29. Klein, M.; Schmidt, A.; Lauer, R. Ontology-centred design of an ambient middleware for assisted living: The case of soprano. Towards Ambient Intelligence: Methods for Cooperating Ensembles in Ubiquitous Environments (AIM-CU). Proceedings of the 30th Annual German Conference on Artificial Intelligence (KI 2007), Osnabrück, Germany, 10–13 September 2007.
30. Tazari, M.R.; Furfari, F.; Valero, Á.F.; Hanke, S.; Höftberger, O.; Kehagias, D.; Mosmondor, M.; Wichert, R.; Wolf, P. The universAAL Reference Model for AAL. In *Handbook on Ambient Assisted Living: Technology for Healthcare, Rehabilitation and Well-Being*; IOS Press: Amsterdam, The Netherlands, 2012; Volume 11, pp. 610–625.
31. The PostgreSQL Global Development Group, PostgreSQL 9.3.3 Documentation. Available online: http://www.postgresql.org/docs/9.3/static/index.html (accessed on 6 January 2014).
32. Silberschatz, A.; Korth, H.F.; Sudarshan, S. *Database System Concepts*; McGraw-Hill: New York, NY, USA, 2011; Volume 6.

33. Paton, N.W.; Diaz, O. Active database systems. *ACM Comput. Surv. (CSUR)* 1999, *31*, 63–103.

34. De Morais, W.O.; Wickström, N. A "Smart Bedroom" as an Active Database System. Proceedings of the IEEE 9th International Conference on Intelligent Environments (IE), Athens, Greece, 18–19 July 2013; pp. 250–253.

35. The PostGIS Development Group, PostGIS 2.1.2dev Manual. Available online: http://postgis.net/docs/manual-2.1/ (accessed on 6 January 2014).

36. Hellerstein, J.M.; Ré, C.; Schoppmann, F.; Wang, D.Z.; Fratkin, E.; Gorajek, A.; Ng, K.S.; Welton, C.; Feng, X.; Li, K.; *et al.* The MADlib analytics library: Or MAD skills, the SQL. *Proc. VLDB Endow.* 2012, *5*, 1700–1711.

37. Cohen, J.; Dolan, B.; Dunlap, M.; Hellerstein, J.M.; Welton, C. MAD skills: New analysis practices for big data. *Proc. VLDB Endow.* 2009, *2*, 1481–1492.

38. Ordonez, C. Data set preprocessing and transformation in a database system. *Intell. Data Anal.* 2011, *15*, 613–631.

39. Sarawagi, S.; Thomas, S.; Agrawal, R. Integrating association rule mining with relational database systems: Alternatives and implications. *Data Min. Knowl. Discov.* 2000, *4*, 89–125.

40. Ordonez, C. Integrating K-means clustering with a relational DBMS using SQL. *IEEE Trans. Knowl. Data Eng.* 2006, *18*, 188–201.

41. Ordonez, C. Building statistical models and scoring with UDFs. Proceedings of the 2007 ACM SIGMOD International Conference on Management of Data, Beijing, China, 12–14 June 2007; pp. 1005–1016.

42. Ordonez, C. Statistical model computation with UDFs. *IEEE Trans. Knowl. Data Eng.* 2010, *22*, 1752–1765.

43. Kim, J.E.; Boulos, G.; Yackovich, J.; Barth, T.; Beckel, C.; Mosse, D. Seamless Integration of Heterogeneous Devices and Access Control in Smart Homes. Proceedings of the 8th International Conference on Intelligent Environments (IE), Guanajuato, Mexico, 26–29 June 2012; pp. 206–213.

44. Dey, A.K.; Abowd, G.D.; Salber, D. A conceptual framework and a toolkit for supporting the rapid prototyping of context-aware applications. *Hum. Comput. Interact.* 2001, *16*, 97–166.

45. Aztiria, A.; Augusto, J.C.; Basagoiti, R.; Izaguirre, A. Accurate Temporal Relationships in Sequences of User Behaviours in Intelligent Environments. In *Ambient Intelligence and Future Trends-International Symposium on Ambient Intelligence (ISAmI 2010)*; Augusto, J., Corchado, J., Novais, P., Analide, C., Eds.; Springer-Berlin: Heidelberg, Germany, 2010; Volume 72, pp. 19–27.

46. Señor, I.C.; Alemán, J.L.F.; Toval, A. Personal Health Records: New Means to Safely Handle Health Data? *Computer* 2012, *45*, 27–33.

47. Otsu, N. A Threshold Selection Method from Gray-Level Histograms. *IEEE Trans. Syst. Man Cybernet.* 1979, *9*, 62–66.

48. Agrawal, R.; Srikant, R. Mining sequential patterns. Proceedings of the Eleventh International Conference on Data Engineering, Taipei, Taiwan, 6–10 March 1995; pp. 3–14.

49. Guralnik, V.; Haigh, K.Z. Learning Models of Human Behaviour with Sequential Patterns. Proceedings of the AAAI-02 Workshop "Automation as Caregiver", Edmonton, Canada, 29 July 2002; pp. 24–30.

50. Agrawal, R.; Imielinski, T.; Swami, A. Mining association rules between sets of items in large databases. *ACM SIGMOD Record* 1993, *22*, 207–216.

51. De Morais, W.O.; Mayr, M.; Wickström, N.; Philippsen, R. Ambient Intelligence and Robotics: Complementing one another to support Ambient Assisted Living. Intelligent Robot Assistants IRAS 2014. Proceedings of the Workshop and Tutorials IAS-13, 13th International Conference on Intelligent Autonomous Systems, Padova, Italy, 15–19 July 2014.

Chapter 3

Review of Potential Characterization Techniques in Approaching Energy and Sustainability

David J. LePoire

Environmental Science Division, Argonne National Laboratory, 9700 S Cass Ave, Argonne, IL 60439, USA.

ABSTRACT

Societal prosperity is linked to sustainable energy and a healthy environment. However, tough global challenges include increased demand for fossil fuels, while approaching peak oil production and uncertainty in the environmental impacts of energy generation. Recently, energy use was identified as a major component of economic productivity, along with capital and labor. Other environmental resources and impacts may be nearing environmental thresholds, as indicated by nine planetary environmental boundaries, many of which are linked to energy production and use. Foresight techniques could be applied to guide future actions which include emphasis on (1) energy efficiency to bridge the transition to a renewable energy economy; (2) continued research, development, and assessment of new technologies; (3) improved understanding of environment impacts including natural capital use and degradation; (4) exploration of GDP alternative measures that include both economic production and environmental impacts; and (5) international cooperation and awareness of longer-term opportunities and their associated potential scenarios. Examples from the U.S. and the international community illustrate challenges and potential.

INTRODUCTION: THE IMPORTANCE OF ENERGY

Energy in the 20th century enabled rapid growth, accelerated technology development, and freedom from many menial tasks. However, the competition for energy contributed to global conflicts, environmental impacts, and large uncertainty in energy prices leading to instabilities. What will energy enable in this century? One perspective is that competition will increase, environmental impacts will be more hazardous, and markets might be more volatile. Hopefully, this scenario will not be realized. Yet, much work needs to be done to ensure the path to sustainable energy use. Ideas from others that might help guide along this path are collected in this paper and placed in a simple framework.

This simple framework is based on the sustainability characteristics of efficiency, robustness, and resilience which are similar to those of sustainable ecosystems [1]. This introduction briefly discusses these issues as integrated economic and environmental efficiency, maintaining an environment which is effective and resilient, and incorporating techniques into decisions to maintain a robust system that does not collapse as many smaller historical societies have done. The paper then proceeds with a discussion of the need of an energy bridge while information is gathered through research and development about the potential sustainable technologies and approaches. The remaining parts of the paper addresses potential tools for each of the sustainability characteristics: (1) technology development to deliver cost-effective energy with reduced environmental impact: (2) understanding and decision tools to support environmental assessments; and (3) exploration of new indices that include both economic and environmental aspects which would help guide the system to a more sustainable state.

Energy is important to sustain lifestyles, organization, and safety. Energy does not determine what we do; it does, however, determine what we can do. What happens when electricity is not available is seen when major power outages occur. In those situations, fundamental services such as food storage, store operations, safety systems, and transportation can quickly deteriorate. While inexpensive energy has been available for decades, significant trends in energy use, energy production, and energy infrastructure indicate major potential future challenges. These challenges include a larger global population leading more energy-intensive lifestyles, diminishing non-renewable fossil fuels, uncertainty regarding the environmental impacts of energy generation, and aging of energy infrastructures. Many countries and regions have prioritized goals to address these challenges by developing inexpensive, reliable, sustainable energy independence with minimal environmental impacts. However, challenges are made more difficult by the scope and interdependence of the issues.

The developed world's lifestyle has significantly changed through the application of new technology and fossil fuel energy sources over the past 150 years. This technology has led to rapid changes in the way we work, move, communicate, heal, and relax. For example, the change in work settings has led to a large increase in economic productivity through the application of relatively inexpensive energy, along with automation and computer networks. This rapid rise in economic productivity is being replicated throughout the world at the same time that the population is increasing. Higher population, along with a higher quality of living is increasing the pressure on limited natural resources [2, 3]. For example, while the demand for energy has been increasing, the amount of fossil fuels seems limited, with about half of the energy already extracted. Its future scarcity will tend to drive prices higher. Fortunately, sufficient fossil fuels have been stored for millions of years and have enabled previous significant advances in science and technology, in addition to searches for new renewable energy sources.

Robert Ayres [4] recently defined sustainability economics as an extension of environmental economics along with the issues of maintaining development while reducing environmental impacts with an emphasis on the linked problems of energy, global climate change, and fossil fuels. He outlines progress towards sustainability through three phases of eco-efficiency. The first phase uses treatment of the pollutants, e.g., reducing emissions from automobiles. These emission reductions regulations reach a marginal benefit compared to other techniques such as increasing the efficiency of making products, e.g., more efficient car production with new and less materials. This phase reaches a limit when the efficiency of the service delivery becomes more productive, e.g., how best to provide the services that a car currently is needed.

Energy use and production also impact the environment. These impacts include particulates and hazardous materials from production and burning, resource and land usage, and water demands to convert heat to useful electricity. The importance of properly managing environmental impacts from energy usage is not new. Historically, progress is made by increasing energy usage with a respect for environmental issues [5]. These issues have included pressure on land resources, soil renewal, sanitary maintenance, deforestation, soot, and acid rain. In addition to the direct environmental impacts, the uneven fossil fuel distribution has led to conflicts and instabilities in both resource-rich and poor regions.

By definition, civilization must return to a sustainable energy economy that works within the environment. This eventual sustainable energy solution will have the advantage of the knowledge gained over the past 200 years to more effectively extract renewable energy from the sun or nuclear resources. Essentially, this transition from one energy sustainable civilization to another, with the intervening few hundreds of years of fossil fuels is a major experiment. Humans have never gone so far from an equilibrium situation. While life itself is far from equilibrium, the current case is doubly so—using energy at a rapid rate and getting that energy from an unsustainable source. Several historical societies have faced similar, although smaller-scale issues. The tie among economic production, energy, and the environment seems to have contributed to breakdowns in many civilizations before, including the Easter Island society, Mesopotamia, the Maya, and the Roman Empire.

In these cases, the energy sources were animal- and human-based and were dependent on the agricultural productivity in increasingly urban conditions. When these civilizations failed, there were others to take their place, find alternative solutions, and move on. In our current situation, we have moved to a more leveraged position with the use of non-renewable fuels, pushed the limits of the global environment, and become dependent on a high-technology lifestyle. Researchers such as Joseph Tainter, Homer-Dixon, and Diamond describe civilizations that might have collapsed with a large influencing factor being the neglect of overall resource use and negative marginal use of resources leading to environmental degradation [5, 6, 7, 8, 9, 10].

The path to energy sustainability requires improvements in many areas such as (1) identifying renewable energy sources that technology is able to economically convert to useful energy without much environmental impact; and (2) reducing the amount of energy needed through improved efficiency in both conversion, e.g., to electricity, and use. Many ideas have been proposed but much work remains to research and develop the technology for a wide-scale deployment. Complicating issues arise through (1) increased energy demand due to simultaneous population growth and global development; (2) the long lead times and capital investments necessary for research, development, deployment; and (3) the coupled nonlinear nature of energy, the economy, and the environment as depicted in Figure 1. This figures shows on the left a conceptual early growth system where energy, environmental, and economic issues can be treated somewhat independently. As they grow to approach natural limits, the components can no longer be treated independently, and a change in one can cause large effects in others. It seems like the current situation is approaching a situation similar to the one on the right with limits of fossil fuels, the potential environmental impacts, and economic effects of growth. These issues of energy, environment, and economics (social system) were highlighted in two of the seven core research questions identified for sustainability science [11]:

- "How can today's operational systems for monitoring and reporting on environmental and social conditions be integrated or extended to provide more useful guidance for efforts to navigate a transition towards sustainability";

- "How can today's relatively independent activities of research planning, monitoring, assessment, anddecision support be better integrated into systems for adaptive management and soicietal learning".

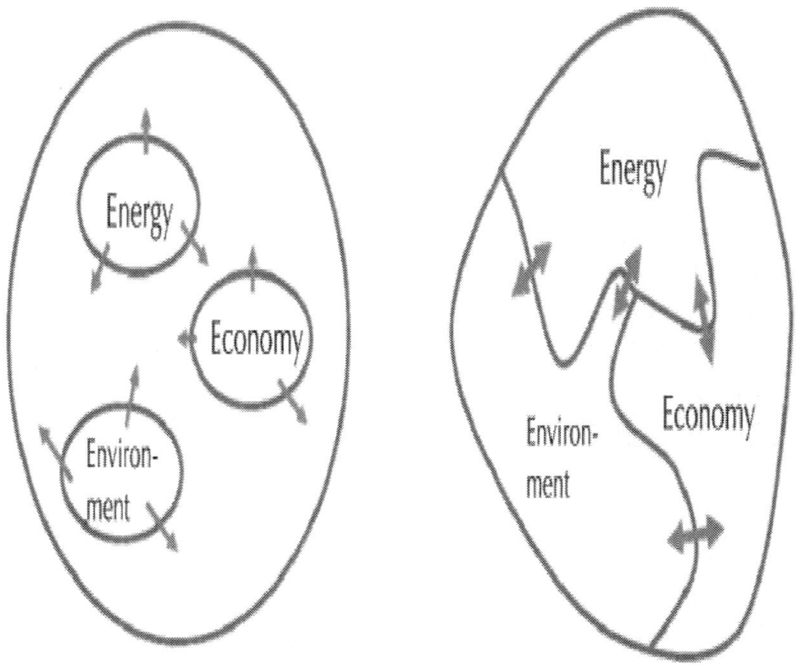

Figure 1: The -diagram shows a conceptual early growth system (left) and a later nonlinear system (right) after growth approaches limits.

Since much research, development, and deployment must be done over a long period, an energy bridge is necessary as the sustainable technologies are tested and either accepted, rejected, or modified. Fortunately, there are still major gains to be realized in efficient use and generation of energy [12, 13]. While not energy sustainable, the substitution of natural gas for coal generation of electricity has led to reduced amounts of carbon dioxide being emitted, which is a step in the right direction for environmental sustainability. Based on life-cycle analysis of production, distribution, and conversion, a 45% reduction in carbon dioxide emission from natural gas use compared to coal [14]. However, since methane is a more potent climate change gas, care is needed to ensure minimal leakage from these activities.

During the bridge period, decisions at all levels will need to be made concerning the viability and sustainability of the system of technologies. The recent debate over the efficacy of using corn-based ethanol as a substitute for gasoline demonstrates the complexity of the issues. This paper identifies several tools that might facilitate measurement, analysis, and understanding in the three areas of energy, environment, and economy [15].

ENERGY BRIDGE

Currently, there is no consensus solution. Energy experts recommend a range of potential paths, including continuation of fossil fuels, nuclear, wind, improvement in energy efficiency, and the development of new technologies [13,16]. Economic viability is a major criterion for an energy solution. There is insufficient knowledge to determine the technology viability, environmental impacts, and economic implications of many new technologies over a long enough period at large-scale deployment. Since this knowledge, research, and experience take time to gather and implement, it seems that a bridge is needed to link current energy efficiency with newer fossil fuel extraction and use methods. The measurement of energy return on energy invested is one tool that is helpful in constructing this bridge. For example, corn-based ethanol production requires a large amount of energy input for the fertilizer, mechanical farming equipment, transportation, and processing. The energy extracted compared to the energy invested in this production is almost equal. Another example is the investment in equipment to increase efficiency. In hybrid cars, the battery is expensive and heavy with more complicated controls. At what point is the investment in this equipment energy cost effective? This bridge give us some time to solidify our understanding and processes, to apply foresight techniques, and develop long-term solutions [17]. The scope and timing means that this is not an isolated problem to be solved independently with engineering certainty but rather requires new views, new collaborations, and planning methods that come with significant uncertainty.

While fossil fuels will probably dominate as a source of energy for another generation, much has been done recently to increase their efficient use [13]. In fact, it seems like energy efficiency has been growing as a substitute for raw energy. This continues the pattern of substituting different fuel sources and technologies (wood, coal, oil, and natural gas) to generate energy from the mid-1800s. The efficiency trend started to gain traction in the 1970s when two oil price shocks hit the economy. Many businesses, governments, and people realized that funds could be better invested in saving energy rather than continual normal use. The energy used now is almost half what it would have been without those efforts [18]. Significant progress could be made in capturing waste heat and combined electricity and heat generation as many countries such as Japan have demonstrated [19].

However, efficiency will only go so far. Eventually, the energy must come from other sources that include solar, wind, hydro, and nuclear. Uncertainties still exist with regard to each energy source's viability in terms of economics, capacity, and integration into the power system. Often a change in one system component is dependent on the upgrade of another. For example, the renewable energy sources that are dependent on weather conditions would be more productive if reliable transmission, storage, and distribution were improved. These systems, however, require much infrastructure development and present transmission or technological obstacles, such as the large storage systems required to implement a viable system.

POTENTIAL FORESIGHT AND ANALYSIS TECHNIQUES

This tightly coupled system of energy, environment, and economic issues involves a wide range of interested groups over a long period of time under significant uncertainty [20, 21, 22]. Issues with this set of conditions, sometimes referred to as "wicked problems", often require trial solutions that are monitored and then modified since no detailed solution can be predetermined. The feedback and iteration improves the solution as uncertainty is reduced. This type of approach requires many foresight techniques such as multiple scenario development, identifying and measuring progress indices, updating flexible road maps, and the inclusion of multiple perspectives and objectives [23]. While foresight techniques have often been applied in defense strategies, the importance of these complicated issues in energy, environment, and the economy suggest that these techniques could be beneficial. Many countries and regions are applying foresight techniques to decision making [24, 25]. Recent projects involving foresight techniques in the United States include the National Intelligence Council's Global Roadmap [26] and a framework to incorporate such techniques at many governance levels [27]. The following sections look at each component (energy, environment, and economy) to identify scenarios, measurements, flexible roadmap tools, and new organization and collaborations [28, 29, 30].

Energy Technology

Energy transition decisions require consideration of many foresight techniques such as (1) scenario construction; (2) redefining the roles of governments, private industry, and collaborations; (3) approaching the problem as an integrated system; (4) continuing assessment with scanning for surprises; (5) development of tools to incorporate uncertainty; and (6) investigating new ways to measure resource use.

Scenarios include those from government, industry, and think tanks, which often include assumptions that range from business as usual, rapid transformation with efficiency gains and renewable sources, and muddling through with a slower development and transition [13]. A major question is the level of government investment in basic research and development. While investments have been made in basic science and technology research, such as the national laboratories and universities, connections with industry have been growing, such as the Energy Innovation Hubs, in the areas of building energy efficiency, solar fuels, energy storage, nuclear design, and critical materials [31]. Direct investment by the government in start-up companies has had some difficulties as the market for electrical transportation and renewable energy generation did not grow as expected after the 2008 recession. Historically, the defense market has provided a bridge from early development to public commercialization. In fact, the defense sector has recognized the larger cost of energy in deployments and in many respects is leading the way in utilizing new energy technology [32].

Revamping the energy systems is a daunting task. Many of the components are interconnected, such as the ability to generate, convert, transport, and use energy. Each requires quite a large investment in infrastructure such as smart grids, renewables that are capital intensive, transportation networks, and integration. These developments require coordination to match the energy characteristics to the reliability and varying demand. Many projects investigate how to transition from the current system to an alternative one; however, most agree the transition will take decades to accomplish [33]. For example, renewables cannot supply much more than 20% before large improvements are made in transmission and storage capabilities.

In addition to systems transitions and integration, the basic components of energy are being developed and assessed. While the U.S. traditionally has not focused on long-term energy policy, some lessons are being learned from abroad and incorporated into its approach of energy technology research. For example, the U.S. Department of Energy (DOE) issued its first quadrennial review of energy technologies in 2012 [31]. Many developed and developing countries contribute to energy related research [34]. There could be many surprises along the way, such as fracking technology, with the potential to expand natural gas and oil development but with uncertain environmental impacts. Other areas of potential surprise include the application of laser drilling, insight into the nature of deep natural gas, techniques to develop methane hydrates, and the potential for broader, safer development of nuclear energy.

A recent Nature paper [35] written by the (then) U.S. Secretary of Energy Steven Chu and the head of the U.S. DOE Advanced Research Projects Agency-Energy (ARPA-E), Arun Majundar, explored opportunities and challenges for a sustainable energy future. They argue for the need for another Industrial revolution to replace the fossil fuel dependency along with energy conservation and efficiency. A path to this goal is the development of cost-effective replacements for fossil fuels accomplished through research and development of new technologies along with the cost reductions through learning during deployment. Great success has been accomplish through the U.S. DOE Sunshot initiative to reduce the cost of solar power to $1 per installed watt for utility capacity of which 40% cover permits and installation. Another possible area for efficiency is in the design of lighter cars and trucks using new materials and improved computer-assisted designs

of internal combustion engines. The development of alternative liquid transportation fuels include approaches such as genetic engineering and artificial photosynthesis supported in the Electrofuels program of E-ARPA and the DOE hub, Joint Center for Artificial Photosynthesis, led at CalTech. Analysis by Bloomberg New Energy Finance of recently completed energy projects showed that onshore wind energy and nuclear were competitive when comparing the unsubsidized levelized cost. While Asia is expanding its nuclear power capacity, setbacks have occurred in the developed countries, specifically Japan and Germany. A possible way to reduce the financial risks of licensing and construction delays is the development of factory manufactured small nuclear reactors. Another promising technology for energy efficiency is using Brayton supercritical CO_2 cycle. to convert heat from power plants into electricity. The efficiency of conversion might approach a 50% conversion efficiency compared to the current average conversion efficiency of generators of about 33%.

One tool to help understand both the trends and their uncertainty is real options analysis [36]. This analysis is similar to the net present value tool but also has the ability to consider possible futures under uncertainty. A number of authors have applied this technique to a range of energy research and development cases, including those for solar power satellites, thorium-based nuclear power, and federal renewable energy research. The key for these tools is to have a good estimate for the scenarios, then track the progress of the research and cost reduction through experience.

Another tool is to refine the measurement of energy. Exergy is an energy measurement that includes the quality of the input and used energy of a process. Energy in the form of slightly heated water is low quality, whereas higher-temperature steam is high quality, and electricity is highest. These tools can be applied to industrial processes to determine the energy use of an industrial ecosystem [37]. Basing economics on such energy and thermodynamics concepts was begun in the 1970s by Georgescu-Roegen and continued by others, including Herman Daly [38]. Recent economic analysis has identified this useful energy measure as a major impact in determining economic productivity [39,40]. Recent analysis which integrates energy and sustainable economics suggests the fundamental importance of energy as a factor in the economy and may explain the "technological factor" seen in 20th century economic growth. However, there are major challenges to implementing a useful sustainable measure to guide decisions. Ayers sees a sequence of phases towards eco-efficiency. First is the efficiency of treatment of pollutants addressed by mitigating the effects of emissions, addressed by many developed countires national environmental laws in the early 1970s. Next is the efficiency of production which reduces the amount of materials and energy required for the same production. Such efficiency was started to be realized after the 1970 oil prices shocks. Finally there is the issue of what really has to be produced to deliver an outcome, *i.e.*, efficiency of service delivery. Some electrical generating plants and states such as California have begun to explore the possibilities in this phase.

Environmental Impacts

Environmental impacts include those that affect air quality, food, water, disease, and land use. A determinant in many of these is the climate, including temperature and precipitation; however, climate is also affected by the rate of soil loss and the application and subsequent pollution of fertilizers and chemical pesticides [41]. There is a concern that widespread economic growth has led to unsustainable resource consumption and use, including fisheries, lumber, energy, soil, and freshwater. An attempt to outline basic environmental measures was made in 2009, with the nine planetary environmental boundaries which include natural resource use (land and water), atmospheric disturbances (ozone, climate change, and aerosols), and releases impacting biological activity and diversity (ocean acidity, chemical pollution, phosphorus, and nitrogen release). Estimates were made for each boundary's natural level, the current level, and the level at which impacts might rapidly rise. Some of these already have impacts much greater than their boundary action level, while a few others have greater uncertainty with undefined boundaries [42].

Environmental scenarios include greater climate change leading to flooding, droughts, and crop changes; biodiversity collapse leading to reduced ecosystem services such as water filtering and pollination; and widespread disease due to rapid contact and migration of species into new climate areas. In addition to these basic resources, pressure has been mounting concerning mineral resources, such as rare earth elements. These are currently frequently used in green energy technologies such as solar cells, energy-efficient electronics, magnets for motors, and advanced batteries [43, 44, 45].

The combinations of advances in non-fossil fuel sources along with potentially slower growth in China might lead to a greener Chinese energy solution [46]. Some estimate that the recent rapid construction of coal power plants in China might end as demand increase slows due to the slower economy, movement to an a more consumer economy, and reduction in need for large energy intensive infrastructure projects. If this scenario were realized, it might lead to pressure other countries to more quickly adopt cleaner energy. The Chinese are facing issues related to emissions from coal including carbon dioxide particulates, and other pollutants such as mercury. The recently completed Minamata International Treaty on mercury emission highlights the environmental impacts of its emission of which coal power plants are a major contributor. While many developed countries implemented many air and water environmental regulations in the early 1970s, mercury release from some coal plants still emit relatively large amounts.

There has been much discussion about the feasibility and impact of a cost placed on carbon dioxide emissions. Such a cost could act as an insurance against potential environmental damage [47]. The costs would impact the relative costs of carbon emitting fuels with those that do not. While it has been difficult to obtain international or national agreement on a way to implement a carbon cost, some countries and more local states and organizations have developed carbon trading markets. One issue with a non-global agreement is the relative cost advantage in trade. The impact of a carbon price would not only shift the relative prices of energy sources but would also encourage the development of alternative and improved technologies. A similar situation was seen in the 1970s when many countries had higher prices on gasoline fuel for cars because of regional availability and higher taxes. This encouraged the development of more efficient cars in those countries. When the oil prices increased in the U.S. during this period, the more efficient imported cars found a relative advantage compared to less energy efficient domestically produced cars, leading to a large shift in the market share among the global automobile makers. This demonstrates that higher prices, even at a regional level, can encourage technology development that can be advantageously applied in global markets as the costs decrease though innovation, learning, and scaling production.

The U.S. EPA recently proposed new regulations to allow 1100 pounds of carbon dioxide per megawatt-hour which would be similar to a new natural gas driven power plant [48]. The current additional cost for carbon capture and storage (CSS) on new coal plants is estimated to be $0.04 per kilowatt-hour, a substantial increase. Very few coal power plants with CSS have been constructed. One new facility in Kemper County Mississippi has demonstrated a process of converting coal into gas, capturing 60% of the carbon dioxide, and then using it for enhanced oil recovery [49]. It is expected that costs could come down through learning and developing larger scales, so the U.S. DOE has funded eight CCS projects along with funds for loan guarantees, and promoting its programs in ARPA-E to improve novel CCS technologies.

Various government activities are investigating possible approaches to environmental impacts. The preamble of the U.S. National Environmental Policy Act (NEPA) of 1969 already highlights many aspects of sustainability and environmental security: "To declare national policy which will encourage productive and enjoyable harmony between man and his environment; to promote efforts which will prevent or eliminate damage to the environment and biosphere and stimulate the health and welfare of man; to enrich the understanding of the ecological systems and natural resources important to the Nation..." NEPA has led to the generation of environmental impact statements (EISs) for government projects and the formation of the President's Council on Environmental Quality (CEQ). Recent EISs have been prepared for the licensing (or relicensing) of nuclear power plants; the siting of wind and solar plants on government lands, such as federal land in the western United States; and the leasing of Outer Continental Shelf areas for energy development purposes such as offshore oil platforms. Some view the NEPA preamble as suggesting a mechanism to broaden consideration of other government activities [50]. Government agencies have been involved with identifying, refining, and testing new indices that realize environmental impacts on air, water, ecosystems, and land use [25,51].

Realigning the Economic Compass

The conventional gross domestic product (GDP) measure of economic progress has been used to guide businesses, governments, and investors for more than 75 years. However, a measure like this is sometimes similar to using a magnetic compass, usually pretty good but difficult if trying to reach limits like the poles. At some point, the compass will lead the voyage in circles, never getting to the poles. Similarly, the GDP has been useful, but its limitations are now apparent. Perhaps we need a refined compass to ensure future progress.

While the GDP measures market economic activity, there are problems in its ability to account for wider activities such as social well-being, and, in a larger sense, environmental sustainability. In 1968, Robert Kennedy stated: "It (our gross national product) measures everything, in short, except that which makes life worthwhile."

It would be nice to evaluate progress toward effectiveness of actions, policy robustness, and potential actions to help decision makers. Such a system would allow greater guidance to preferred future scenarios. An ideal system is a dream; currently, decisions are guided by imperfect systems of information and indices. These current tools have many limitations, thus alternatives are being identified, explored, and tested in many different settings. However, new indices for guiding decisions are difficult to define, measure, and deploy. Such a system requires numerous measurements, interpretation, models, and validation [24, 52, 53].

Many countries developed national strategies for sustainability after the first Earth Summit in 1992. The United States formed a President's Council on Sustainable Development that developed recommendations in the 1990s. In 1998, an interagency group developed a set of potential alternative (or experimental) indices [51]. In 2008, the French sponsored the Commission on the Measurement of Performance and Social Progress to review alternative economic measures and implementations led by Joseph Stiglitz [24]. In 2010, the British released a report of alternative sustainable indicators. In 2011, the U.S. Government Accountability Office compared various regions' approaches to defining and using alternative indices. Many nongovernmental organizations (NGOs) have formed to advocate for various indicators and uses, including EthicalMarkets, wikiprogress, and Citizens Network for Sustainable Development.

A recent paper [54] continued investigating the Genuine Progress Indicator by extending it to 17 countries representing over half the global population and GDP. The world's GPI/capita trend from the

1950s the present was then estimated. The GPI, similar to the Index of sustainable Economic Welfare (ISEW) proposed by Daly and Cobb in the late 1980s, includes consideration of GDP, volunteering and housework while excluding environmental costs, and costs of crime and income inequality. They found most developed countries had parallel increases of GPI/capita and GDP/capita from the 1950s to the mid-1970s. After that the GDP/capita continued to climb by about 80% up to 2010, while the GPI/capita remained relatively constant or slightly decreasing. Exceptions to this were Japan where the parallel growth continued to about 1990, followed by stalls in both GDP/capita and GPI/capita. China's parallel growth continued to about the mid-1990s, followed by rapid GDP growth and GPI/capita relative constancy. The estimate global GPI/capita rose by a factor of 2 since 1950 to about $4000 (2005 $USD) maintaining a value about 2/3's that of the GDP until 1975. Since then the GDP/capita has risen by almost another factor of 2 while the GPI/capita has remained almost constant with a slight decrease of about 10%.

NEXT STEPS

Richard Smalley, from Rice University and the 1996 Nobel Prize winner in chemistry for his discovery of fullerenes in the mid-1980s, offered a possible way to tie these transitions together to hopefully take advantage of their benefits while managing or at least reducing the possible risks. He suggested that the U.S. should lead an international effort to explore the possibilities of using nanotechnology in the energy sector. He argues that energy is the key to many international problems, such as water, food, environment, poverty, terrorism and war, disease, education, democracy, and population [55].

Various countries are participating in science and technology research for energy and environmental areas [34]. Large projects include the International Thermonuclear Experimental Reactor (ITER), which is exploring magnetically confined fusion energy. Other examples include the sharing of approaches to energy efficiency between China and the United States by Lawrence Berkeley National Laboratory. Collaboration among countries and among various types of organizations, for example, universities, national laboratories, and industry, to search for new solutions ensures that basic research is shared, but it does raise issues regarding maintaining the incentive for funding research, especially at more applied levels. In the case of advanced battery research, the difference between basic and applied research is diminishing. Organizational and collaborative structures are being explored to enhance discovery and motivate greater participation. Each organization has a different prospective and potential benefits with royalties for discovering, profits for producing, jobs for manufacturing location, and energy and environmental benefits from its use [56].

While there has been great progress in understanding and coordinating information, energy, environmental, and economic issues require continual exploration and refinement as they involve new domains in both the time and spatial scales. The interaction among energy, the environment, and the economy can also be likened to the difficult task of squeezing a balloon. Any irregularities in the compression will result in parts of the balloon popping out, just as the way to the integrated solution of this problem will cause difficulties at different points in development. Continual refinement and flexibility in addressing the issues are required.

The knowledge, measures, tools, organizations, trials, and feedback discussed above are just part of the work being done to tackle, frame, and resolve these issues. It is a very interesting and exciting experiment. The process will take a long time, similar to other major transitions in history.

ACKNOWLEDGMENTS

Work supported by the U.S. Department of Energy under Contract No. DE-AC02-06CH171357. The views expressed are those of the author and do not reflect the official policy or position of Argonne, UChicago-Argonne, the University of Chicago, or DOE.

REFERENCES

1. Gunderson, L.H.; Holling, C.S. *Panarchy: Understanding Transformations in Human and Natural Systems*; Island Press: Washington, DC, USA, 2002.

2. Ausubel, J.H. Can technology spare the earth? *Am. Sci.* 1996, *84*, 166–178.
3. LePoire, D. Long-term Population, Productivity, and Energy Use Trends in the Sequence of Leading Capitalist Nations. *Technol. Forecast. Soc. Change* 2010, *77*, 1303–1310.
4. Ayres, R.U. Sustainability economics: Where do we stand? *Ecol. Econ.* 2008, *67*, 281–310.
5. Ponting, C. *A New Green History of the World: The Environment and the Collapse of Great Civilizations*, rev ed.; Penguin Books: New York, NY, USA, 2007.
6. Wright, R. *NonZero: The Logic of Human Destiny*, 1st ed.; Pantheon Books: New York, NY, USA, 2000.
7. Korten, D.C. *The Great Turning: From Empire to Earth Community*; Kumarian Press: San Francisco, CA, USA, 2006.
8. Homer-Dixon, T.F. *The Upside of Down: Catastrophe, Creativity, and the Renewal of Civilization*; Island Press: Washington, DC, USA, 2006.
9. Tainter, J.A. Complexity, Problem Solving, and Sustainable Societies. In *Getting Down to Earth*; Constanza, R., Ed.; Island Press: Washington, DC, USA, 1996.
10. Diamond, J.M. *Collapse: How Societies Choose to Fail or Succeed*; Viking Press: New York, NY, USA, 2005.
11. Kates, R.W.; Clark, W.C.; Corell, R.; Hall, M.J.; Jaeger, C.C.; Lowe, I.; McCarthy, J.J.; Schellnhuber, H.J.; Bolin, B.; Diskson, N.M.; *et al*. Sustainability Science. *Science* 2001, *292*, 641–642.
12. Granade, H.; Creyts, J.; Derkach, A.; Farese, P.; Nyquist, S.; Ostrowski, K. Unlocking energy efficiency in the US economy. Available online: http://www.mckinsey.com/client_service/electric_power_and_natural_gas/latest_thinking/unlocking_energy_efficiency_in_the_us_economy (accessed on 16 March 2014).
13. Lovins, A.B.; the Rocky Mountain Institute. *Reinventing Fire: Bold Business Solutions for the New Energy Era*; Chelsea Green Publishing: White River Junction, VT, USA, 2011.
14. Evans, A.; Strezov, V.; Evans, T.J. Assessment of sustainability indicators for renewable energy technologies. *Renew. Sust. Energ. Rev.* 2009, *13*, 1082–1088.

15. Ness, B.; Urbel-Piirsalu, E.; Anderberg, S.; Olsson, L. Categorising tools for sustainability assessment. *Ecol. Econ.* 2007, *60*, 498–508.
16. Heaberlin, S.W. *A Case for Nuclear-Generated Electricity, Or, Why I Think Nuclear Power Is Cool and Why It Is Important That You Think So Too*; Battelle Press: Columbus, OH, USA, 2004.
17. Ayres, R.U.; Ayres, E. *Crossing the Energy Divide: Moving from Fossil Fuel Dependence to A Clean-Energy Future*; Wharton School Publishing: Upper Saddle River, NJ, USA, 2010.
18. Devezas, T.; LePoire, D.; Matias, J.C.O.; Silva, A.M.P. Energy scenarios: Toward a new energy paradigm. *Futures* 2008, *40*, 1–16.
19. Heinberg, R. *The End of Growth: Adapting to Our New Economic Reality*; New Society Publishers: Gabriola Island, BC, Canada, 2011.
20. Bradshaw, G.A.; Borchers, J.G. Uncertainty as Information: Narrowing the Science-policy Gap. *Ecol. Soc.* 2000, 4. Article 7. Availab e online: http://www.ecologyandsociety.org/vol4/iss1/art7/ (accessed on 16 March 2014).
21. Conklin, E.J. *Dialogue Mapping: Building Shared Understanding of Wicked Problems*; Wiley: Chichester, UK; Hoboken, NJ, USA, 2006.
22. World Economic Forum. Global Risks Report 2013. Available online: http://reports.weforum.org/global-risks-2013/ (accessed on 16 March 2014).
23. LePoire, D.J. *Threading the Environmental Needle: Applying New Tools to Reduce Uncertainty in Environmental Foresight, in Sustainable Futures, Strategies and Techniques*; Wagner, C.G., Ed.; World Future Society: Bethesda, MD, USA, 2010.
24. Stiglitz, J.E.; Sen, A.; Fitoussi, J.-P. Report by the Commission on the Measurement of Economic Performance and Social Progress. Available online: http://www.stiglitz-sen-fitoussi.fr/documents/rapport_anglais.pdf (accessed on 16 March 2014).
25. U.S. Government Accountability Office. Key Indicator Systems: Experiences of Other National and Subnational Systems Offer Insights for the United States. Available online: http://www.gao.gov/products/GAO-11-396 (accessed on 16 March 2014).

26. Global Trends 2030: Alternative Worlds. U.S. National Intelligence Council 2012. Available online: http://globaltrends2030.files.wordpress.com/2012/11/global-trends-2030-november2012.pdf (accessed on 16 March 2014).
27. Fuerth, L.S.; Faber, E.M.H. Anticipatory Governance Practical Upgrades. Available online:http://www.wilsoncenter.org/sites/default/files/Anticipatory_Governance_Practical_Upgrades.pdf (accessed on 16 March 2014).
28. American Energy Innovation Council. A Business Plan for America's EnergyFuture. 2010. Available online: http://americanenergyinnovation.org/wp-content/uploads/2012/04/AEIC_The_Business_Plan_2010.pdf (accessed on 16 March 2014).
29. Henderson, H. Inside the Global Green Transition. Available online: http://www.ethicalmarkets.com/wp-content/uploads/2012/03/2012-HH-Inside-the-Global-Green-Transition.pdf (accessed on 16 March 2014).
30. Lempert, R.J.; Popper, S.W.; Bankes, S.C. *Shaping the Next One Hundred Years: New Methods For Quantitative, Long-Term Policy Analysis*; RAND: Santa Monica, CA, USA, 2003.
31. U.S. Department of Energy. Science for Our Nation's Energy Future: Transforming the Future of Energy and the Environment. Available online: http://www.energyfrontier.us/sites/all/themes/basic/pdfs/EFRC-Brochure_051911_Front.pdf (accessed on 16 March 2014).
32. Alic, J.; Sarewitz, D.; Thernstrom, S.; Doom, T. Energy Innovation at the Department of Defense: Assessing the Opportunities. Available online: http://bipartisanpolicy.org/library/report/energy-innovation-department-defense-assessing-opportunities (accessed on 16 March 2014).
33. President's Council of Advisors on Science and Technology (PCAST). Report to the President on Accelerating the Pace of Change in Energy Technologies Through an Integrated Federal Energy Policy. Available online: http://www.whitehouse.gov/sites/default/files/microsites/ostp/pcast-energy-tech-report.pdf (accessed on 16 March 2014).
34. U.S. National Research Council. S&T Strategies of Six Countries: Implications for the United States. Available online: http://www.

nap.edu/catalog.php?record_id=12920 (accessed on 16 March 2014).

35. Chu, S.; Majumda, A. Opportunities and Challenges for a sustainable energy future. *Nature* 2012, *488*, 294–303.
36. Siddiqui, S.; Marnay, C.; Wiser, R.H. Real Options Valuation of US Federal Renewable Energy Research, Development, Demonstration, and Development. In Proceedings of the 10th Annual Power Research Conference on Electricity Industry Restructuring, Berkeley, CA, USA, 18 March 2005.
37. Connelly, L.; Koshland, C.P. Exergy and Industrial Ecology. *Exergy Int. J.* 2001, *1*, 146–165.
38. Daly, H.E. *Beyond Growth: The Economics of Sustainable Development*; Beacon Press: Boston, MA, USA, 1996.
39. Ayres, R.U. Energy and economic growth. In Proceedings of the Sustainable Energy Production and Consumption: Benefits, Strategies and Environmental Costing, Naples, Italy, 4–7 July 2007.
40. Jackson, J.T.R. *Occupy World Street: A global Roadmap For Radical Economic and Political Reform*; Chelsea Green Publishing: White River Junction, VT, USA, 2012.
41. Brown, L.R. *Plan B: Rescuing a Planet Under Stress and A Civilization In Trouble*, 1st ed.; W. W. Norton & Company: New York, NY, USA, 2003.
42. Rockstrom, J.; Steffen, W.; Noone, K.; Persson, A.; Chapin, F.S.; Lambin, E.; Lenton, T.M.; Scheffer, M.; Folke, C.; Schellnhuber, H.J.; *et al.* Planetary Boundaries: Exploring the Safe Operating Space for Humanity. *Ecol. Soc.* 2009, 14. Article 32. Available online: http://www.ecologyandsociety.org/vol14/iss2/art32/ (accessed on 16 March 2014).
43. Hecker, J.H. Peace and Sustainable Development through Environmental Security. Available online: http://www.envirosecurity.org/espa/PDF/IES_ESPA_ESA_Methodology.pdf (accessed on 16 March 2014).
44. Steinbruner, J.D.; Stern, P.C.; Husbands, J.L. *Climate and Social Stress: Implications for Security Analysis*; National Research Council: Washington, DC, USA, 2012.

45. Matthew, R.A. Is Climate Change a National Security Issue? *Issues Sci. Technol.* 2011, 27. Available online: http://www.issues.org/27.3/matthew.html (accessed on 16 March 2014).
46. China Council for International Cooperation on Environment and Development. China's Pathway towards a Low Carbon Economy. 2009. Available online: http://www.cciced.net/encciced/policyresearch/report/201205/P020120529358137604609.pdf (accessed on 16 March 2014).
47. Yohe, G.; Andronova, N.; Schlesinger, M. To Hedge or Not Against an Uncertain Climate Future? *Science* 2004, *306*, 416–417.
48. Biello, D. How to Win the War on Coal. *Scientific American*, 18 December 2013.
49. U.S. Government Printing Office. The Future of Coal: Utilizing America's Abundant Energy Resources. Available online: http://www.gpo.gov/fdsys/pkg/CHRG-113hhrg82223/pdf/CHRG-113hhrg82223.pdf (accessed on 16 March 2014).
50. Dernbach, J.C. *Stumbling Toward Sustainability*; Environmental Law Institute: Washington, DC, USA, 2002.
51. U.S. Interagency Working Group on Sustainable Development Indicators. Sustainbable Development in the United States: An Experimental Set of Indicators. Available online: http://teclim.ufba.br/jsf/indicadores/SDI%20US%20SUST%20DEVEL%20INDICAT.PDF (accessed on 16 March 2014).
52. Haggart, B. The Gross Domestic Product and Alternative Economic and Social Indicators. Available online: http://publications.gc.ca/Collection-R/LoPBdP/BP/prb0022-e.htm (accessed on 16 March 2014).
53. U.K. Department for Environment Food and Rural Affairs. Measuring progress: Sustainable development indicators 2010. Available online: http://archive.defra.gov.uk/sustainable/government/progress/documents/SDI2010_001.pdf (accessed on 16 March 2014).
54. Kubiszewski, I.; Costanza, R.; Franco, C.; Lawn, P.; Talberth, J.; Jackson, T.; Aylmer, C. Beyond GDP: Measuring and achieving global Genuine Progress. *Ecol. Econ.* 2013, *93*, 57–68.

55. Smalley, R. Our Energy Challenge. In Processdings of the Energy & Nanotechnology Conference, Rice University, Houston, TX, USA; 2003.

56. Lowe, M.; Tokuoka, S.; Trigg, T.; Gereffi, G. Lithium-ion Batteries for Electric Vehicles: The US Value Chain. Available online: http://www.cggc.duke.edu/pdfs/Lowe_Lithium-Ion_Batteries_CGGC_10-05-10_revised.pdf (accessed on 16 March 2014).

Chapter 4

Carbon Nanofiber Concrete for Damage Detection of Infrastructure

Y.L. Mo[1], and Rachel Howser Roberts[1]

[1]University of Houston, USA

INTRODUCTION

Fiber research in concrete construction is an ongoing field and the use of carbon nanofibers (CNF) is examined. Fibers improve brittle materials such as concrete by enhancing tensile strength, ductility, toughness, and conductivity. Short-fiber composites are a class of strain sensor based on the concept of short electrically conducting fiber pull-out that accompanies slight and reversible crack opening. For a fiber composite to have strain sensing ability, the fibers must be more conducting than the matrix in which they are embedded, of diameter smaller than the crack length, and well dispersed. Their orientations can be random, and they do not have to touch one another. The electrical conductivity of the fibers enables the direct current (DC) electrical resistivity of the composites to change in response to strain change or temperature, allowing sensing.

NANOTECHNOLOGY IN CONCRETE

Despite the fact that nanotechnology is a relatively recent development in scientific research, the introduction of the concept is credited to Nobel Prize winner Richard Feynman from his 1959 lecture, "There's Plenty of Room at the Bottom" [1]. Feynman considered the possibility of direct manipulation of individual atoms as a powerful form of synthetic chemistry. Decades later, Feynman's concept morphed into the field of nanotechnology. According to the National Science Foundation and National Nanotechnology Initiative, the definition of nanotechnology includes three elements [2]:

- The size range of the material structures under consideration should be approximately 100 nanometers;
- The nanotechnology should have the ability to measure or transform at the nanoscale;
- There should be properties that are specific to the nanoscale as compared to the macro or micro scale.

Following this definition, in the past 25 years nanotechnology has expanded from Feynman's idea and now finds applications in fields ranging from medical devices to nano-reinforced concrete [3, 4].

To date, the awareness and application of nanotechnology in the construction industry are increasing; however, progress is uneven in the current early stages of its practical exploitation. Bartos [5] presents three reasons for this phenomenon:

- The nature of the construction industry differs greatly from the other industries doing research in nanotechnology. The final products coming from the construction industry are not mass-produced and require relatively long service lives, differentiating it from the products from the microelectronics, information technology, and automotive industries.

- Historically, there is a very low level of investment in construction research and development.
- Research in nano-related research and development requires very high initial capital investment

Despite these difficulties, there have been significant advances in nanoscience of cementitious materials with an increase in the understanding of basic phenomena in cement at the nanoscale. These include structure and mechanical properties of the hydrate phases, origins of cement cohesion, cement hydration, interfaces in concrete, and mechanisms of degradation [6]. A major nanotechnology application is to include nano-sized reinforcement in cement-based materials such as carbon nanotubes or nanofibers.

FIBER REINFORCED CONCRETE

Concrete, composed of fine and coarse aggregates held together by a hydrated cement binder, is one of the most important construction materials and is used in diverse project areas including house foundations, high rise tower components, highways, and dams. Hydrated cement is a brittle material that is an order of magnitude stronger in compression than in tension. To compensate for this weakness reinforcement consisting typically of rebar or fibers are added to the concrete.

The use of fibers to reinforce brittle materials can be traced back to ancient times when straw and hair was added to mud bricks. The modern development of the use of fibers in construction began in the 1960s with the addition of steel fibers to reinforced concrete structures. This was closely followed by the addition of polymeric fibers, glass fibers, and carbon fibers in the 1970s, 80s, and 90s, respectively [7].

Fibers improve brittle materials such as concrete by enhancing tensile strength, ductility, toughness, and conductivity [8-13]. Fibers are typically used in two forms: short randomly dispersed fibers in a cementitious matrix or a continuous mesh of fibers used in thin sheets. Here we will focus on randomly dispersed fibers used to arrest cracks. The cracking process within concrete begins with the onset of isolated nanocracks. These nanocracks grow together to form localized microcracks, which in turn grow together to form macrocracks. These macrocracks widen to form cracks visible with the naked eye. Fibers arrest these cracks by forming bridges across them. With increasing tensile stress, a bond failure eventually occurs, and the fiber will pull out of the concrete allowing the crack to widen.Fig. 1 shows the bridging action of fibers across micro and macrocracks in concrete.

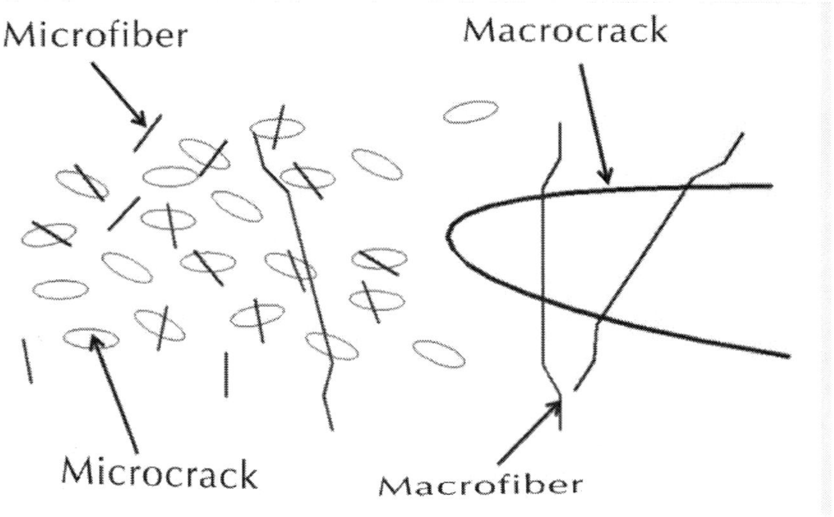

Figure 1: Bridging action of fibers across micro and macrocracks.

NANOREINFORCEMENT IN CEMENT-BASED MATERIALS

Since the discovery of carbon nanotubes (CNT) in 1991 [14], researchers have desired to implement the unique mechanical, thermal, and electronic properties of CNT and CNF in cement-based composites. Single-wall CNT (SWCNT), multi-wall CNT (MWCNT) and CNF are graphene ring-based materials with aspect ratios greater than 1000 and high surface areas [6, 15]. CNT and CNF have moduli of elasticity in the range of terrapascals and tensile strength on the order of gigapascal [6, 16,17]. SWCNT consist of a single graphene sheet wrapped into a seamless cylinder, while, as the name suggests, MWCNT inhere of multiple concentric sheets of graphene wrapped around a hollow core. CNF are cylindric nanostructures with graphene layers arranged as stacked cones, cups, or plates. CNF are adventagious because their stacked structure presents exposed edge planes not present in CNT that intoduce increased surface area and better bond characteristics. Because of their structure, CNF are easier to produce and cost 100 times less than SWCNT [18]. Because of the increased bond surface and lower cost, CNF are more attactive than CNT for application in cement-based composites.

CNT AND CNF DISPERSION

The majority of nanoreinforced composite research has been completed on polymers containing CNT or CNF [6, 19, 20]. One of the main reasons for this is because uniform dispersion is difficult in cement-based materials. Well dispersed CNF results in uniform calcium-silicate-hydrate (CSH) gel formation, which improves the structural and electrical properties of the concrete [21]. CNT and CNF are inherently hydrophobic and are attracted due to Van der Waals forces, causing the fibers to tend to agglomerate hindering their dispersion in solvants [17, 22-24].

Several solutions have been proposed to solve this issue including dispersing the fibers through milling, ultrasonication, high shear flow, elongational flow, functionalization, in addition to surfactant and chemical dispersement systems [24]. These methods primarily fall into two categories: mechanical and chemical dispersion. The mechanical dispersion methods, such as ultrasonification, while effective in seperating the fibers, can fracture them decreasing their aspect ratio. Chemical methods use surfactants or functionalization to make the fibers less hydrophobic, reducing their tendancy to agglomerate. However, many of the chemicals used can digest the fibers causing the fibers to become less effective. The surfactants also often cause bubbles to form in the composite negatively effecting the strength of the material.

Gao et al [12] proposed a dispersion method specifically used for CNF/CNT dispersion in cement-based materials that eliminates the beforementioned drawbacks. In this method, a high-range water reducer (HRWR) is used to create a self-consolidating concrete (SCC). ACI Committee 237 Self-Consolidating Concrete offers the following definition for SCC [25]:

> Self-consolidating concrete (SCC) is highly flowable, non-segregating concrete that can spread into place, fill the formwork, and encapsulate the reinforcement without any mechanical consolidation.

SCC is a product of technological advancements in the area of under-water concrete technology where the mixtures must ensure high fluidity and high resistance to washout and segregation. Okamura originally advocated SCC in 1986, and the first success with SCC occurred in 1988 [26]. The use of SCC has gained wide acceptance for savings in labor costs, shortened construction time, a better finish, and an improved work environment [27-30].

Advancement in SCC technology was primarily possible due to the introduction of new chemical admixtures that improved and controlled the SCC rheological properties. Better performing SCC mixes were produced on the advent of melamine, naphthalene, polycarboxylate, and acrylic based HRWR superplasticizers and viscosity modifying agents (VMA).

Gao et al [12] proposed using SCC because acceptable SCC is not only highly flowable, but it is also highly stable and homogenious on a macro scale. The Prestressed Concrete Institute (PCI) stipulates the following criteria for SCC [26]:

- Filling ability – The property that determines how fast SCC flows under its own weight and completely fills intricate spaces with obstacles, such as reinforcement, without losing its stability.
- Passing ability – the ability of SCC to pass through congested reinforcement and adhere to it without application of external energy.
- Stability – the ability of SCC to remain homogenous by resisting segregation, bleeding, and air popping during transport and placing as well as after placement.

Gao et al [12] studied SCC containing CNF to see if the same effect was present on the nano scale. In Gao et al's mixing procedure, HRWR, water, and CNF are mixed in a laboratory-grade blender. Simultaneously, fine aggregate, course aggregate, and cement are combined in a centrifugal mixer. The CNF mixture is then slowly added to the mixer to gain a homogenous mix. The fresh concrete was used to create cylinders that were tested in compression. After the test, pieces of the cylinders were observed under a scanning electron microscope (SEM). The SEM showed significant CNF clumping in specimens made of normal CNF concrete and uniform distribution in SCC containing CNF, as shown in Figs. 2 and 3, respectively.

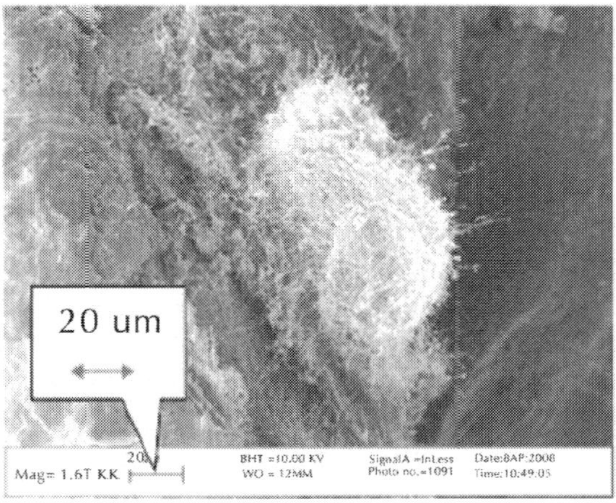

Figure 2: Scanning Electron Microscope Image of CNF Clump in Normal Cement (1670x Magnification).

Figure 3: Scanning Electron Microscope Image of Well Dispersed CNF in a Uniform Self-Consolidating Cement (9410x Magnification).

STRAIN SENSING ABILITY OF CNT/ CNF CEMENT-BASED MATERIALS

Smart materials are materials that sense their environment and respond to changes in strain, temperature, moisture, pH, and/or electric or magnetic fields. CNT/CNF composites qualify as smart materials since they can be used to measure strain and temperature [4, 13, 32-35]. There are two types of strain sensing, reversible and irreversible. The measurement of irreversible strain allows structural health monitoring, while the sensing of reversible strain permits dynamic load monitoring. Structural health monitoring is the process of implementing a damage detection and characterization strategy for engineering structures. Dynamic load monitoring can detect loads on structures as they are applied and removed in real time. These are important technologies because they gauge the ability of a structure to perform its intended function despite aging, degradation, or disasters. Typically, monitoring reversible strain is more difficult because it can only be monitored in real time. Additionally, reversible strain tends to be smaller than irreversible strain [31].

Strain sensing refers to the ability to measure an electrical or optical response corresponding to a strain. Chen and Chung [31] give the following requirements for a structural sensor:

- Wide strain/stress range of detection (from small strains up to failure)
- Response being reversible upon stimulus removal (necessary for repeated use of the sensor)
- Ease of measuring the response (without the need of expensive peripheral equipment)
- Presence of the sensor having no bad effect on the structural properties of the structure
- Chemical stability ǎnd durability
- Low cost

Current commonly used strain sensors include strain gages, fiber optic sensors, and piezoelectric sensors, which all suffer from high cost, poor durability, and the need for expensive peripheral equipment including electronics and lasers. Because of this, the use of sensors in civil structures is uncommon [31]. CNT/CNF composites could become a better option as a strain sensor because however, technology may provide a way to make them more cheaply in the future.

CNT and CNF cement-based materials exhibits properties necessary for reversible strain monitoring and electromagnetic interference (EMI) shielding. Short-fiber composites were found to be a class of strain sensor based on the concept of short electrically conducting fiber pull-out that accompanies slight and reversible crack opening. For a CNT/CNF composite to have strain sensing ability, the fibers must be more conducting than the matrix in which they are embedded, of diameter smaller than the crack length, and well dispersed. Their orientations can be random, and they do not have to touch one another [32, 33]. The electrical conductivity of the fibers enables the DC electrical resistivity of the composites to change in response to strain damage or temperature, allowing sensing [13, 32-35].

CARBON FIBER CEMENT AND MORTAR SELF-SENSING APPLICATIONS

Around the same time that CNT were discovered, reseachers were adding carbon microfibers to cement-based materials and studying their mechanical properties. In 1992 while studying the mechanical properties of carbon microfibers dispersed in mortar, Yang and Chung [35] noted that the electrical resitivity of mortar containing these fibers dramatically decreased by up to several orders of magnitude.

This idea of electrically conducting concrete led to Chen and Chung proposing an intrinsically smart concrete containing carbon microfibers [8]. Chen and Chung prepared mortar cubes containing carbon microfibers and tested them cyclically. They discovered that the electrical resistivity of the concrete increased irreversibly upon compressive loading up to approximately 1/3 the compressive strength of the mortar. After this point, the resistance reversibly increased and decreased upon loading and unloading of the specimens. Chen and Chung concluded that carbon fiber reinforced concrete can serve as a smart structural material. Chen and Chung followed this experiment with a more detailed cyclic experiment on carbon fiber mortar under cyclic loads [31]. After this test, they concluded that the initial irreversible behavior is due to permanent damage associated with the fiber/matrix interface weakening. They attributed the reversible behavior to crack opening with fiber pull-out and crack closing with fibers pushing back in.

CNT are the most conductive fibers presently known and are, therefore, more ideal for electrical applications than their micro-scale counterparts [36, 37]. CNT and CNF are also attractive for use in cement-based composites because of strength and high aspect ratios [6, 16, 17]. Li et al proposed adding MWCNT to mortar for improved mechanical properties [14]. Li et al confirmed that the flexural and compressive strength of the concrete was enhanced, but they did not study the electrical properties. The same group later studied the electrical volume resistivity of cement paste containing CNT measured using the four-probe method [39]. They applied a cyclic compressive load to a 40.0 mm by 40.0 mm by 160.0 mm (1.575 in. by 1.575 in. by 6.30 in.) rectangular prism made of the material. The fractional change in the volume resistivity oscillated up to approximately 10% with the oscillation of the compressive load.

DAMAGE DETECTION OF CNF CONCRETE COLUMNS

Gao et al expanded the work on self-sensing cement-based materials by studying 152.4 mm by 305 mm (6.00 in. by 12.00 in.) cylindars made of concrete containing CNF [12]. Gao et al crushed the cylinders monotonically and studied the electrical resistance variation. They observed electrical resistance variations up to 80% and concluded that concrete containing CNF can be used for self-structural health monitoring.

Howser et al continued Gao et al's work and extended it to a full scale reinforced concrete column containing CNF [4, 12]. A self-consolidating CNF concrete (SCCNFC) column was built and tested under a reversed cyclic load. The structural behavior and the self-sensing ability were examined. The results were compared to the structural and self-sensing ability of a traditional self-consolidating reinforced concrete (SCRC) and a self-consolidating steel fiber concrete (SCSFC) specimen.

All of the columns were 508 mm (20.0 in.) tall with cross-sections of 305 mm by 305 mm (12.00 in. by 12.00 in.). Each specimen contained six #8 (25.4 mm or 1.00 in. diameter) rebar, which corresponded to 3.27% logitudinal steel by volume of concrete. The SCRC and SCCNFC columns contained #2 stirrups with a spacing of 120.7 mm (4.75 in.) providing transverse reinforcement of 0.287% by volume of concrete. Since the columns were designed to be shear critical, the maximum reinforcement spacing was chosen based on the ACI 318 specifications [25]. See Fig. 4 for the cross-section used for the SCRC and SCCNFC columns. SCSFC column contained no transverse reinforcement, as shown in Fig. 5. Each of the columns was rigidly connected to similar foundations. See Fig. 6 for the elevation view of the SCRC and SCCNFC columns and foundations. The SCSFC column is identical to that shown in Fig. 6, except it does not contain transverse reinforcement. Fig. 7shows the experimental set-up.

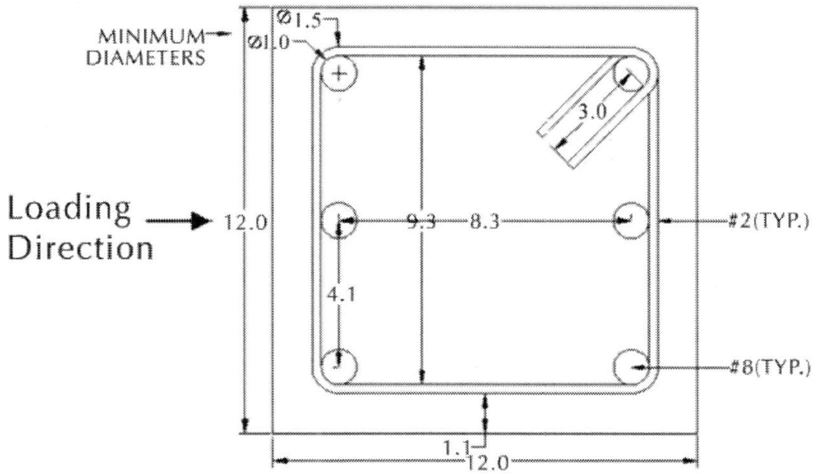

Figure 4: Cross-Section of SCRC and SCCNFC Columns (dimensions in inches).

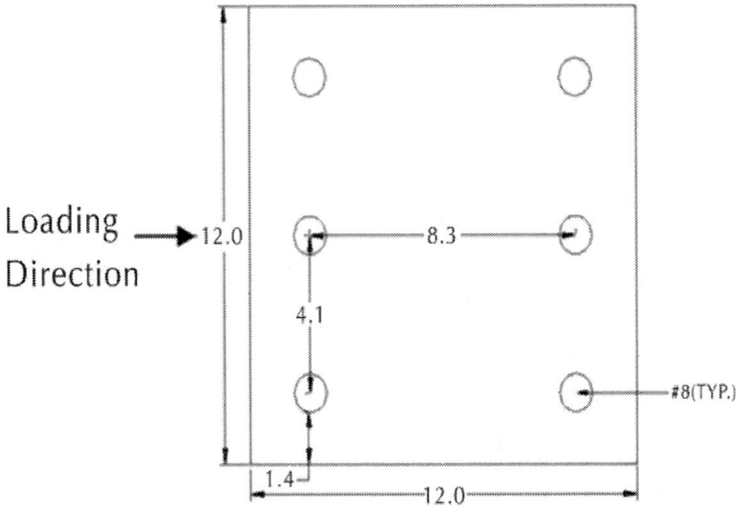

Figure 5: Cross-Section of SCSFC Column (dimensions in inches).

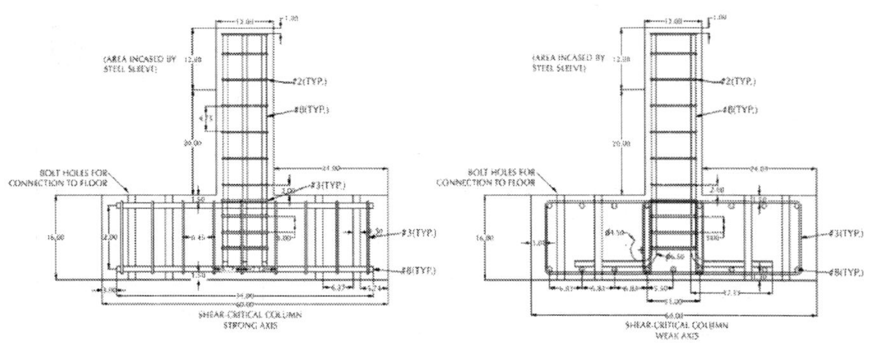

Figure 6: Elevation View of the Strong Axis of the Shear-Critical SCRC and SCCNFC Columns and Foundations (dimensions in inches).

Figure 7: Experimental Set-Up.

The properties of the materials used for the three mixes were as follows:

a. Cement: The cement used in all mixtures was ASTM Type III Portland cement.
b. Fly Ash: Class C fly ash was used for the SCSFC mix and Class F fly ash was used for the SCRC mix.
c. Coarse Aggregate: Crushed limestore with a maximum diameter of ¾″ was used in the SCCNFC column. River rock with a maximum diameter of ¾″ was used in the other columns.
d. Fine Aggregate: Natural river sand with a fineness modulus of 2.71 was used in all mixes.
e. High Range Water Reducer (HRWR): Glenium® 3200HES was used in the SCCNFC column and Glenium® 3400 HES was used

in the other columns. Both chemicals were polycarboxylate admixtures from BASF Chemical Co.

f. Viscosity Modifying Agent (VMA): RHEOMAC® VMA 450 was used in the specimens and also supplied by BASF Chemical Co.

g. Steel Fibers: Dramix® ZP305 fibers were used in the SCSFC mix. This was a hooked fiber with a specific gravity of 7.85. The diameter of the fiber is 0.55 mm (0.0217 in.) and the length is 30 mm (1.18 in.) resulting in an aspect ratio of 55.

h. Carbon Nanofibers: Pyrograf Products, Inc. PR-19-XT-LHT-OX fibers were used in this study. The specific gravity of the fibers was 0.0742. The diameter of the fibers was 149 nm (5.87e-6 in.) and the length was 19 μm (7.48e-4 in.) resulting in an aspect ratio of 128.

The mix proportions used for the three columns can be seen in Table 1. One percent fiber by volume was used for both of the fiber columns chosen based on literature review. It was discovered by Gao et al that CNF has an optimal dosage of approximately 1% by volume [12][12]. It was found by many researchers that increased steel fiber increases concrete properties; however, after a percentage of 1% fibers by volume, the concrete becomes increasingly less workable, which could cause problems in construction such as honeycombing [39-41].

Table 1: Mix Proportions in kg/m^3 (lb/yd^3) of Concrete

Material	**SCRC Mix**	**SCSFC Mix**	**SCCNFC Mix**
Cement	446 (752)	446 (752)	457 (771)
Fly Ash (Class C)	-	299 (504)	-
Fly Ash (Class F)	299 (504)	-	-
Fine Aggregate	937 (1580)	937 (1580)	898 (1514)
Coarse Aggregate (Limestone)	-	-	859 (1448)
Coarse Aggregate (River Rock)	491 (827)	491 (827)	-
Water	224 (377)	224 (377)	182 (307)
Glenium® 3400HES	2.81 (4.73)	2.81 (4.73)	-
Glenium® 7700HES	-	-	2.34 (3.94)
REHEOMAC® VMA 450	5.69 (9.59)	5.69 (9.59)	-
Steel Fibers	-	79.8 (134)	-
Carbon Nanofibers	-	-	3.23 (5.45)

The main goal of testing the SCCNFC column was to prove that concrete containing CNF can be used as a sensor. To test the electrical properties of the concrete, wire meshes were constructed and embedded in each of the columns. The wire meshes were made of 12.7 mm (1/2 in.) hardware cloth with 14 gauge copper wire soldered to it. The wire extended outside of the column. The four probe method for measuring resistance was implemented, and the meshes were placed in the column as shown in Fig. 8. A power supply was attached to the top mesh that provided a current of approximately 31 V DC. An ammeter was attached to the bottom mesh and connected back to the power supply to complete a circuit. The current measured by the ammeter was recorded continuously during the tests by hand. Additional voltmeters were attached to the two middle meshes on both the north and south sides of the column to measure voltage. The voltage readings were also recorded continuously throughout the test.

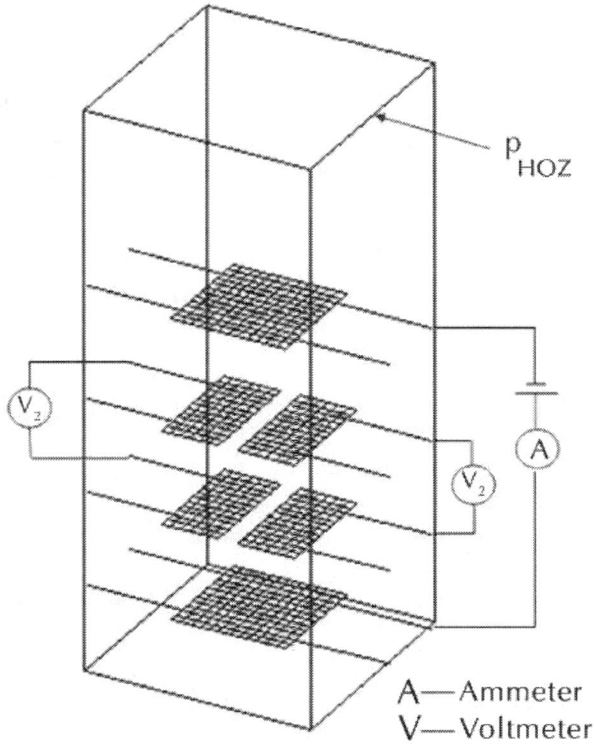

Figure 8: Four Probe Method of Resistance Measurement.

The first step of the load program was to apply an axial load that would remain constant during the course of the test. The axial load equaled one-tenth of each of the columns calculated axial capacity. The axial capacity is dependent on the compressive strength of the concrete, so the axial force for each specimen varied.

After the application of the axial load, a reversed-cyclic load was added using a 649 kN (146 kip) capacity actuator. The intended load path was to use force control to complete two cycles each of ±89 kN (20 k), ±178 kN (40 k), and ±267 kN (60 k). A positive force denotes a push by the actuator while a negative force represents a pull. At the point of longitudinal steel yielding, the test switched to displacement control and completed two cycles each at a displacement ductility of 2, 3, 4, etc. Once failure occurred, a descending branch on the load versus displacement curve was obtained in displacement control mode.

The load path followed for the SCRC column specimen can be seen in Fig. 9 with the first cracks, switch to displacement control and failure marked. The first crack on the south side of the column occurred at -178 kN (-40 k). The first shear crack formed on the column during the first -178 kN (-40 k) cycle at -178 kN (-40 k) on the west side. The column failed in shear and crushing of concrete at 276 kN (62 k). The west side of the column exhibited crushing of the concrete struts with large shear cracks. The east side exhibited local crushing at the actuator connection. The maximum displacement at the top of the column (drift) was 12.7 mm (0.50 in.).

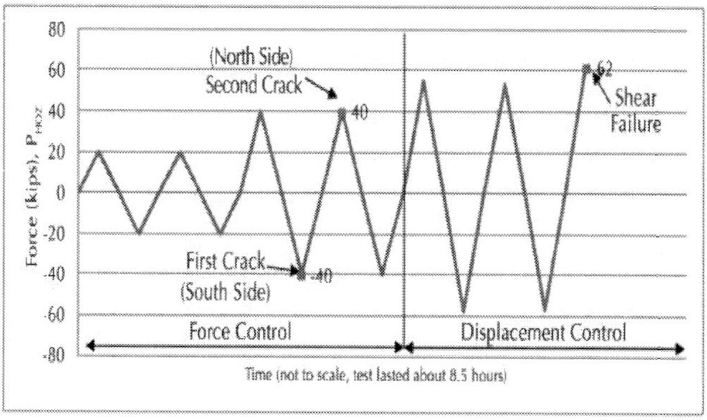

Figure 9: SCRC Column Load Path.

The load path followed for the SCSFC column can be seen in Fig. 10 with the first cracks and failure marked. The first shear and flexural cracks formed on the column during the second 178 kN (40 k) cycle at 178 kN (40 k) on the west and north sides, respectively. The second flexural crack formed on the south side during the second -178 kN (-40 k) cycle at -178 kN (-40 k). The column failed suddenly in shear and crushing at 347 kN (78.0 k) on the west side of the column before the rebar yielded. The maximum displacement was 8.38 mm (0.33 in.).

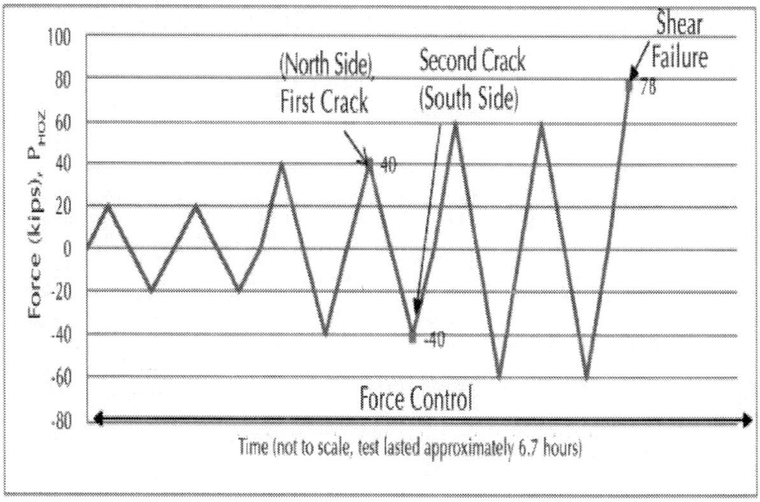

Figure 10: SCSFC Column Load Path.

The actual load path followed for the SCCNFC column can be seen in Fig. 11. The pump shut down during the test, and the actuator unloaded during the fifth cycle of the test. The pump was turned back on and the test resumed. The first flexural crack formed on the column at 160 kN (36 k) on the east, west and north sides. The second flexural crack formed on the east, west and south sides at a load of -158 kN (-35.6 k). The column failed in the combined modes of shear and concrete crushing due to flexure at 298 kN (67 k) on the west side of the column. The maximum displacement was 10.16 mm (0.4 in.).

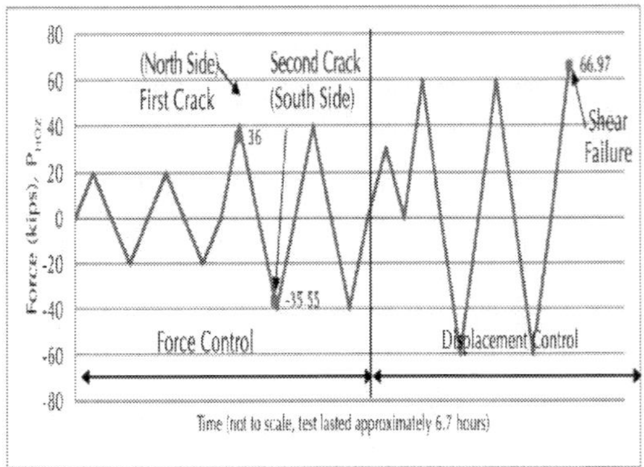

Figure 11: SCCNFC Column Load Path.

During each of the column tests, the electrical resistance was determined to check the self-sensing ability of the concrete. The electrical readings showed a great correlation between the peaks in the applied horizontal force, strain, and resistance plots for the SCCNFC column but little correlation between the resistance plots and the force or strain plots for the SCRC or SCSFC column. Fig. 12shows the relationship between the SCRC column's horizontal force, LVDT strain, and electrical resistance versus time on the north side of the column. There is no relationship between the peaks and valleys in the electrical resistance and the load or strain on the north side of the column.

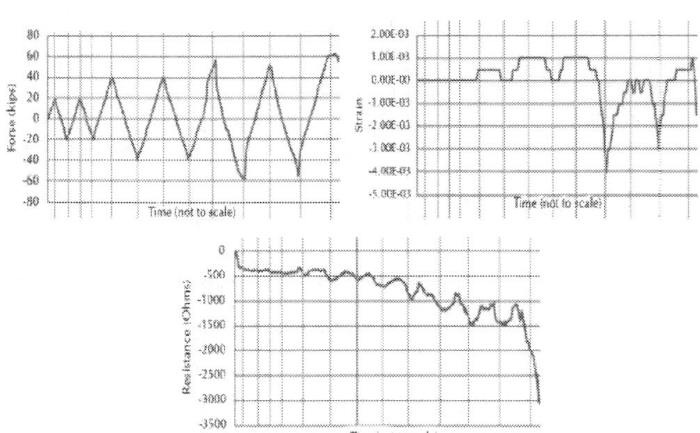

Figure 12: SCRC Column Comparison of Horizontal Force, LVDT Strain and Electrical Resistance on North Side.

Fig. 13 shows the SCSFC column's force, strain, and resistance versus time on the north and south sides of the column, respectively. As shown by the grey vertical lines, there is not a relationship between the peaks and valleys in the resistance and load or strain until major cracking began to occur. After major cracking began to occur, the peaks and valleys in the electrical resistance began to correspond with the load and strain peaks and valleys. This point is shown by the dashed line in Fig. 13.

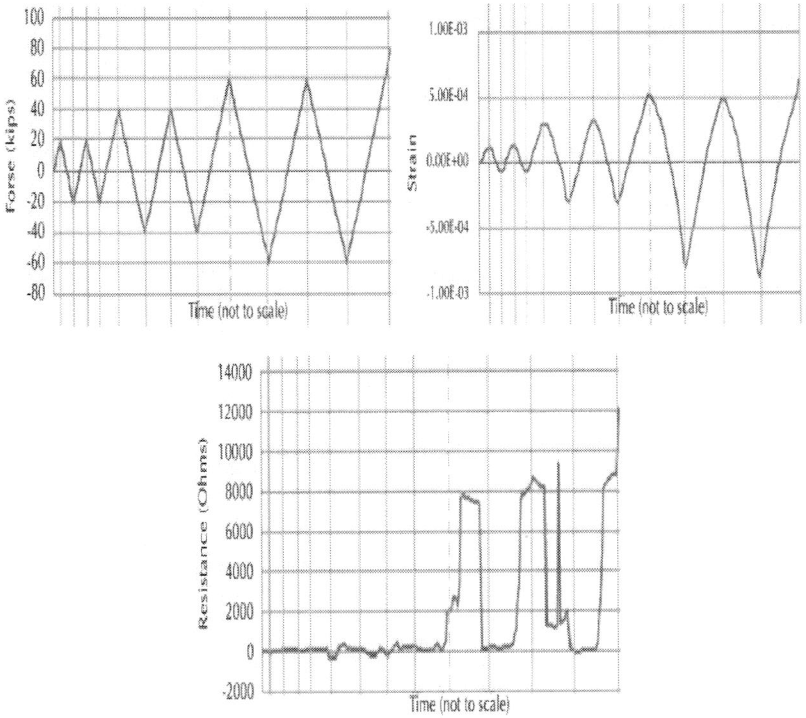

Figure 13: SCRC Column Comparison of Horizontal Force, LVDT Strain and Electrical Resistance on North Side.

Fig. 14 shows relationship between the SCCNFC column's horizontal load, LVDT strain and electrical resistance versus time on the north side of column. As shown by the vertical lines in the grid, there is very good correlation between the force, strain and resistance. On the north side of the column, the peaks and valleys matched up until the point that the column was greatly damaged.

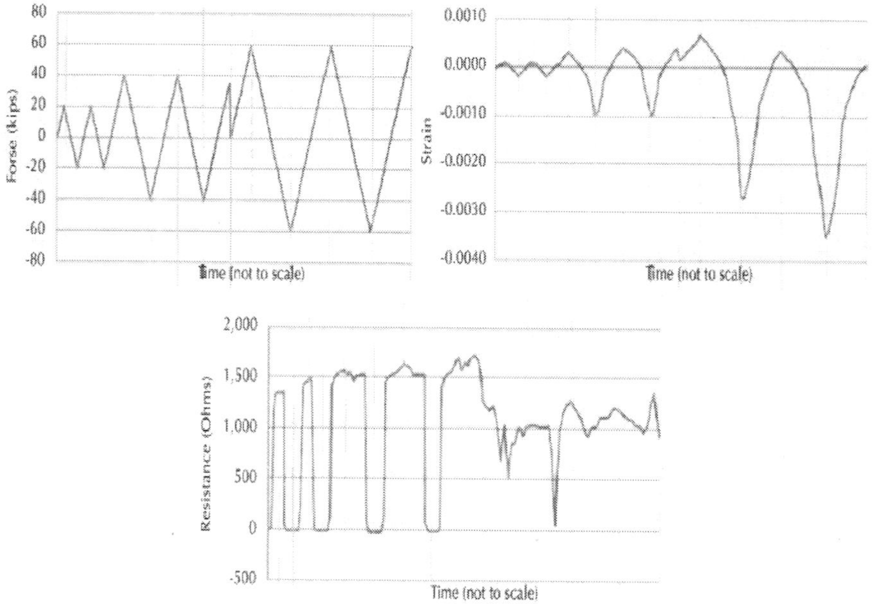

Figure 14: SCCNFC Column Comparison of Horizontal Force, LVDT Strain and Electrical Resistance on North Side.

Because of the strong correlation found between the horizontal load, LVDT strain, and electrical resistance verses time graphs for the SCCNFC column, the electrical resistance variation (ERV) was calculated and compared to the deflection at the top of the column. ERV is the measured electrical resistance minus the initial electrical resistance quantity divided by the initial electrical resistance. Fig. 15 shows the relationship between the ERV and deflection at the top of the column for the first five cycles of the test. It is obvious from Fig. 15 that the column shows major damage at approximately a deflection of 2.03 mm (0.08 in.). This corresponds to the steel yielding in the SCCNFC column. This proves that SCCNFC can be used as a type of self-structural health monitoring system.

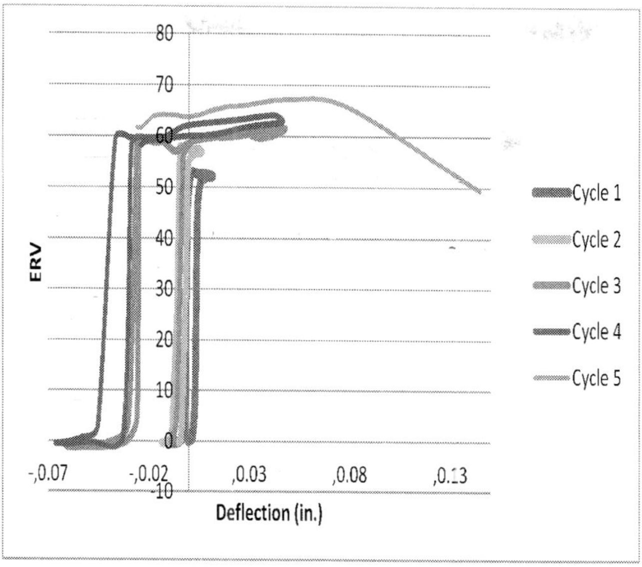

Figure 15: SCCNFC Column ERV versus Horizontal Deflection.

CONCLUSIONS

Self-consolidating carbon nanofiber concrete (SCCNFC) follows the definition for nanotechnology set forth by the National Science Foundation and National Nanotechnology Initiative [2]. The size range of the carbon nanofibers (CNF) is approximately 100 nanometers, the SCCNFC is able to measure damage in the composite, and the CNF have properties that are specific to the nanoscale.

Well-dispersed CNF improves the strength and stiffness of concrete. Excess concentration leads to poorly dispersed CNF clumps inside the concrete and has a negative effect on both strength and electrical sensitivity. Highly workable and stable self-consolidation concrete (SCC) can maintain its workability and stability with the addition of fibers. SCC greatly increases the dispersion of carbon nanofibers (CNF) [12].

As proven by Gao et al [12] and Howser et al [4], SCCNFC can be used as a reversible strain sensor. In Howser et al's test [4], the peaks and valleys in the electrical resistance readings of the SCCNFC match the peaks and valleys of the applied force and the strain in the concrete. While the peaks and valleys in the electrical resistance readings of the self-consolidating reinforced concrete and self-consolidating steel fiber concrete specimens occasionally matched, there was not enough correspondence to safely assume that these concretes could be used as a reversible strain sensor. It was concluded that when an appropriate dosage of CNF is used, SCCNFC can be used for self-structural health monitoring.

REFERENCES

1. Feynman, R. There's Plenty of Room at the Bottom. Engineering and Science 1960; 23, 22-36.
2. Roco, M. C. National Nanotechnology Initiative - Past, Present, Future. In W. A. Goddard, D. Brenner, S. E. Lyshevski, & G. J. Iafrate (Eds.), Handbook of Nanoscience, Engineering, and Technology (2nd ed.). Boca Raton, FL: CRC Press 2007; 3.1-3.26.
3. Narayan, R. J., Kumta, P. N., Sfeir, C., Lee, D.-H., Choi, D., Olton, D. Nanostructured ceramics in medical devices: Applications and prospects. Journal of the Minerals, Metals and Materials Society 2004; 56 (10), 38-43.
4. Howser, R. N., Dhonde, H. B., Mo, Y. L. Self-sensing of carbon nanofiber concrete columns subjected to reversed cyclic loading. Smart Materials and Structures 2011; 20 (8).
5. Bartos, P. Nanotechnology in Construction: A Roadmap for Development. American Concrete Institute Special Publication 2008; 254, 1-14.
6. Sanchez, F., Sobolev, K. Nanotechnology in concrete - A review. Construction and Building Materials 2010; 24, 2060-2071.
7. Shah, S. P., Naaman, A. E. Mechanical Properties of Glass and Steel Fiber Reinforced Mortar. ACI Journal 1976; 73 (1), 50-53.
8. Chen, P.-W., Chung, D. D. Concrete reinforced with up to 0.2 vol% of short carbon fibres. Composites 1993; 24 (1), 33-52.
9. Li, H., Xiao, H.-g., Yuan, J., Ou, J. Microstructure of cement mortar with nano-particles. Composites Part B: Engineering 2004; 35 (2), 185-189.

10. Li, H., Zhang, M.-h., Ou, J.-p. Abrasion resistance on concrete containing nano-particles for pavement. Wear 2006; 260 (11-12), 1262-1266.
11. Li, H., Zhang, M.-h., Ou, J.-p. Flexural fatigue performance of concrete containing nano-particles for pavement. International Journal of Fatigue 2007; 29 (7), 1292-1301.
12. Gao, D., Strum, M., Mo, Y. L. Electrical resistance of carbon-nanofiber concrete. Smart Materials and Structures 2009; 18 (9), 1-7.
13. Iijima, S. Helical microtubules of graphitic carbon. Nature 1991, 354, 56-58.
14. Li, G. Y., Wang, P. M., Zhao, X. Mechanical behavior and microstructre of cement composites incorporating surface-treated multi-walled carbon nanotubes. Carbon 2005; 43, 1239-1245.
15. Salvetat, J. P., Bonard, J. M., Thomson, N. H., Kulik, A. J., Forro, L., Benoit, W., et al. Mechanical properites of carbon naonotubes. Applied Physics A 1999; 69, 255-260.
16. Makar, J. M., Beaudoin, J. J. Carbon nanotubes and their application in the construction industry. 1st International Symposium on Nanotechnology in Construction, Paisley, Scotland 2003; 331-341.
17. Kang, I., Heung, Y. Y., Kim, J. H., Lee, J. W., Gollapudi, R., Subramaniam, S., et al. Introduction to carbon nanotube and nanofiber smart materials. Composites: Part B 2006; 382-394.
18. Coleman, N. N., Khan, U., Blau, W. J., Gun'ko, Y. K. Small but strong: A review of the mechanical properties of carbon nanotube-polymer composites. Carbon 2006; 44, 1624-1652.
19. Makar, J. M., Margeson, J. C., Luh, J. Carbon nanotube/cement composites-early results and potential applications. 3rd International Conference on Construction Materials: Performance, Innovation and Structural Implications. Vancouver, B.C.: NRC Institute for Research in Construction; National Research Council Canada 2005; 1-10.
20. Chung, D. D. Dispersion of Short Fibers in Cement. Journal of Materials in Civil Engineering, ASCE 2005; 17 (4), 379-383.

21. Tzeng, Y., Huang, T. S., Chen, Y. C., Liu, C., Liu, Y. K. Hydration Properties of Carbon Nanotubes and Their Effects on Electrical and Biosensor Applications. New Diamond and Frontier: Carbon Technology 2004; 14 (3), 193-201.
22. Baughman, R. H., Cui, C., Zakhidov, A. A., Iqbal, Z., Barisci, J. N., Spinks, G. M., et al. Carbon Nanotube Actuators. Science 1999; 284 (5418), 1340-1344.
23. Baughman, R. H., Zakhldov, A. A., de Heer, W. A. Carbon Nanotubes--the Route Towards Applications. Science 2002; 297, 787-792.
24. Hilding, J., Grulke, E. A., Zhang, Z. G., & Lockwood, F. Dispersion of Carbon Nanotubes in Liquids. Journal of Dispersion Science and Technology 2003; 24 (1), 1-41.
25. ACI 318. Building Code Requirements for Structural Concrete. Farmington Hills, Michigan: American Concrete Institute; 2008.
26. PCI, TR-6-03. Interim Guidelines for the Use of Self-Consolidating Concrete in Precast/Prestressed Concrete Institute Member Plants. Chicago: Precast/Prestressed Concrete Institute; 2003.
27. Gaimster, R., Foord, C. Self-compacting concrete. Concrete 2000; 34, 23-25.
28. Khayat, K. H., Hu, C., Monty, H. Stability of Self-Consolidating Concrete, Advantages, and Potential Applications. Proceedings of First International RILEM Symposium on Self-Compacting Concrete. Stockholm, Sweden 1999; 143-152.
29. Okamura, H., Ozawa, K. Mix-Design for Self-Compacting Concrete. Concrete Library of the Japanese Society of Civil Engineers 1995; 107-120.
30. Tanaka, K., Sato, K., Watanabe, S., Arima, I., Seunaga, K. Development and Utilization of High-Performance Concrete for the Construction of the Akashi Kaikyo Bridge. High Performance Concrete in Severe Environments, ACI SP-140 1993; 25-51.
31. Chen, P.-W., Chung, D. D. Concrete as a new strain/stress sensor. Composites Part B: Engineering 1996; 27 (1), 11-23.
32. Chung, D. D. Strain sensors based on the electrical resistance change accompanying the reversible pull-out of conducting short fibers in a less conducting matrix. Smart Materials and Structures 1995; 4 (1), 59-61.

33. Chung, D. D. Cement-matrix composites for smart structures. Smart Materials and Structures 2000; 9 (4), 389-401.
34. Xiao, H., Lan, C., Ji, X., Li, H. Mechanical and sensing properties of structural materials with nanophase materials. Pacific Scientific Journal 2003; 5, 11-17.
35. Yang, X., Chung, D. D. Latex-modified cement mortar reinforced by short carbon fibres. Composites 1992; 23 (6), 453-460.
36. Thess, A., Lee, R., Nikolaev, P., Dai, H., Petit, P., Robert, J., et al. Crystalline Ropes of Metallic Carbon Nanotubes. Science 1996; 273, 483-487.
37. Wei, B. Q., Vajtai, R., Ayayan, P. M. Reliability and current carrying capactiy of carbon nanotubes. Applied Physics Letters 2001; 79 (8), 1172-1174.
38. Li, G. Y., Wang, P. M., Zhao, X. Pressure-sensitive properties and microstructure of carbon nanotube reinforced cement composites. Cement & Concrete Composites 2007; 29, 377-382.
39. Narayan, R. J., Kumta, P. N., Sfeir, C., Lee, D.-H., Choi, D., Olton, D. Nanostructured ceramics in medical devices: Applications and prospects. Journal of the Minerals, Metals and Materials Society 2004; 56 (10), 38-43.
40. Padmarajaiah, S. K., Ramaswamy, A. Comparative Study on Flexural Response of Full and Partial Depth Fiber-Reinforced High-Strength Concrete. Journal of Materials in Civil Engineering 2002; 14 (2), 130-136.
41. Aoude, H., Cook, W. D., Mitchell, D. (2009). Behavior of Columns Constructed with Fibers and Self-Consolidating Concrete. ACI Structural Journal 2009; 106 (3), 349-357.

Chapter 5

Applying Lean Thinking in Construction and Performance Improvement

Remon Fayek Aziz, Sherif Mohamed Hafez

Structural Engineering Department, Faculty of Engineering, Alexandria University, Egypt

ABSTRACT

The productivity of the construction industry worldwide has been declining over the past 40 years. One approach for improving the situation is using lean construction. Lean construction results from the application of a new form of production management to construction. Essential features of lean construction include a clear set of objectives for the delivery process, aimed at maximizing performance for the customer at the project level, concurrent design, construction, and the application of project control throughout the life cycle of the project from design to delivery. An increasing number of construction academics and professionals have been storming the ramparts of conventional construction management in an effort to deliver better value to owners while making real profits. As a result, lean-based tools have emerged and have been successfully applied to simple and complex construction projects. In general, lean construction projects are easier to manage, safer, completed sooner, and cost less and are of better quality. Significant research remains to complete the translation to construction of lean thinking in Egypt. This research will discuss principles, methods, and implementation phases of lean construction showing the waste in construction and how it could be minimized. The Last Planner System technique, which is an important application of the lean construction concepts and methodologies and is more prevalent, proved that it could enhance the construction management practices in various aspects. Also, it is intended to develop methodology for process evaluation and define areas for improvement based on lean approach principles.

INTRODUCTION

Since the 1950s, lean production or Toyota production system principles have evolved and were successfully implemented by Toyota Motor Company. Toyota production system had two pillar concepts: (1) Just In Time flow (JIT) and (2) Autonomation (smart automation) as shown in more details in Fig. 1.

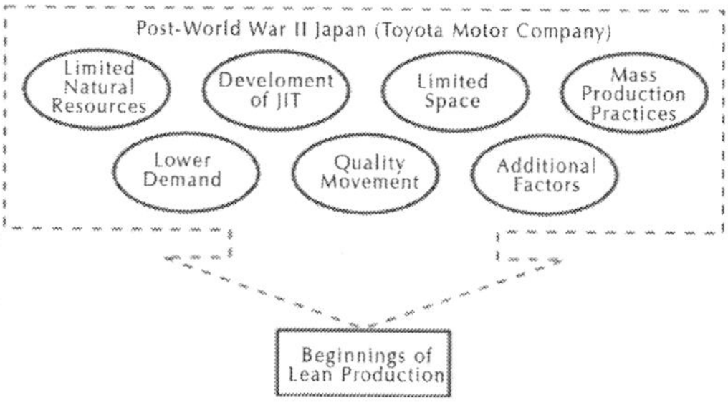

Figure 1: Beginning of lean production [1].

The term "lean" was coined by the research team working on international auto production to reflect both the waste reduction nature of the Toyota production system and to contrast it with craft and mass forms of production [2]. Starting from efforts to reduce machine setup time and influenced by TQM, a simple set of objectives was developed for the design of the production system including to (1) Identify and deliver value to the customer value: eliminate anything that does not add value; (2) Organize production as a continuous flow; (3) Perfect the product and create reliable flow through distributing information and decision making; and (4) Pursue perfection: Deliver on order a product meeting customer requirements with nothing in inventory. Lean production aims to design and make things differentiated from mass and craft forms of production by the objectives and technique, and to optimize performance of the production system against a standard of perfection to meet unique customer requirements. In the beginning of the 1990s, the new production philosophy, which is known by several different names, is as follows: (1) world class manufacturing; (2) lean production; and (3) new production system. This philosophy is the emerging mainstream approach. It is practiced, at least partially, by major manufacturing companies in America and Europe. The new approach has also diffused to new fields, like customized production, services, administration, and product development. Since 1992, Koskela [3] has reported the adaptation of lean production concepts in the construction industry and presented a production management paradigm where production was conceptualized in three complementary ways, namely as (1) Transformation; (2) Flow; and (3) Value generation (TFV) theory of production. This tripartite view of production has led to the birth of lean construction as a discipline that subsumes the transformation-dominated contemporary construction management [4] and [5]. Managing construction under lean is different from typical contemporary practice because it (1) has a clear set of objectives for the delivery process; (2) is aimed at maximizing performance for the customer at the project level; (3) designs concurrently product and process; and (4) applies production control throughout the life of the project. The first goal of lean construction must be to fully understand the physics of production, the effects of dependence and variation along supply and assembly chains. In lean construction as in much of manufacturing, (1) *Planning*: defining criteria for success and producing strategies for achieving and (2) *Control*: causing events to conform to plan, and triggering learning and re-planning are two sides of a coin that keeps revolving throughout a project. In this research, principles, methods, and the implementation phases of lean construction will be discussed showing the waste in construction and how it could be minimized. The Last Planner System (LPS) technique, which is an important application of the lean construction concepts and methodologies and is more prevalent, proved that it could enhance the construction management practices in various aspects and bring numerous advantages, so that the construction projects will be more stable and less stressful for all involved stakeholders by reducing dependencies and variations to identify and eliminate waste (non-value adding activities).

RESEARCH BACKGROUND

Construction management and technology are the two key factors influencing the development of the construction industry. Over the past 40 years, although several new and advanced technologies have been applied to construction projects, the efficiency of the industry has remained quite low [6], [7] and [8]. For example, the productivity of the USA construction industry has been declining since 1964 [9]. A similar decline in construction productivity has also occurred in other countries. Japan, for example, decreased from 3714 to 2731 Yen/Man/Hours over the period of 1990–2004. The main reason for this appears to be that the new technologies cannot effectively reduce the cost of design and construction while, at the same time, improving the management of the construction process. For example, although the Computer Aided Design (CAD) technology has improved the efficiency of drawing, it cannot reduce design errors and these, in turn, can cause the need for rework of construction making it difficult for construction managers to optimize the construction process to reduce cost [10] and [11]. This is a particularly relevant issue for Design/Build (D/B) projects, where the aim is to reduce cost and increase quality by an improved constructability of the building design. However, the new technologies cannot, as yet, effectively support the implementation of D/B projects. Therefore, the application of both appropriate new technology and contemporary management concepts is likely to be two effective approaches to improve construction industry efficiency. One of the new management philosophies that have been considered for the UK construction industry is that of lean thinking [12]. Lean construction, much like current practice, has the goal of better meeting customer needs while using less of everything, a term coined by the International Group for Lean Construction in 1993, Gleeson and Townend [13] had been investigated by many researchers in recent years. This refers to the application of lean production principles and practices in design–construction processes to maximize value and to reduce waste [14] and [15]. Some successful experience in implementing lean construction has been achieved. Conte and Gransberg [16], for example, examined the principles used in applying lean construction by over 20 construction companies in Brazil. Similarly, Wright [17] presented several cases involving the use of lean construction. However, the application of lean construction is still in its initial stages. In order to improve the implementation of lean construction, Miller et al. [18] proposed the harmonization between main contractors and subcontractors as a prerequisite, while Thomas et al.[19] and [20] proposed reducing variability to improve performance and improving labor flow reliability for better productivity as lean construction principles. Bashford et al. [21] also introduced the production management model into residential construction; it was presented a lean management model for the construction of high-rise apartment buildings, which was developed to LEAPCON that did a simulation of lean construction of high-rise apartment buildings, Sacks et al. [22]. Sacks et al. [23] proposed a building information modeling based on lean production system. Gabriel [24] illustrated through case studies, the developed concept of lean project management, which recognized that the system and philosophy are now better understood by participants, who themselves provide and contribute to the project control mechanisms. The project manager becomes an individual client-based leader of the project team, the source of decision making, and the conduit for communication. The approach had been successful on complex public arts buildings over 10 years and could be of general application on wide range of project management and management by project environments. Al-Sudairi et al. [25] reported interim results of a study to evaluate lean principles when applied to construction using computer simulation. Data for a structural steel erection process were modeled to form the experimental tool for evaluating lean principles. In all cases, the simulated principles improved project performance. Performance improved dramatically when all principles are simultaneously applied. Thomas et al. [26] investigated the lean principle, which more reliable flows lead to better labor performance. Actual data from three bridge construction projects are examined to document the instances of poor flow resource reliability and its effect on labor performance. Inefficient labor hours are calculated. The results showed that there was strong support that more reliable material, information, and equipment availability contributes to better performance. Salem et al. [27] tested the effectiveness of some lean construction tools, in particular, those tools that can be applied in medium size construction firms. Due to the success of the lean production system in manufacturing, the construction industry has adapted lean techniques to eliminate waste and increase profit. A field study was conducted to evaluate the effectiveness of some lean construction techniques. The data collection methods included direct observations, interviews, questionnaires, and documentary analysis. The effectiveness of the lean construction tools was evaluated through the lean implementation measurement standard and performance criteria. It was found that last planner, increased visualization, daily huddle meetings, and first-run studies achieved more effective outcomes than expected. However, the results of implementation of 5S process and fail safe for quality did not meet the expectations of the tool champions and the research team. It was found that there is a need for behavioral changes and training for effective use of lean tools. Most of the lean construction tools, selected for the project, are either ready to use or are recommended with some modifications. Agbulos et al. [28]presented development that involved the improved work methods and engineered productivity standards for the various drainage operations. The work measurement concept was implemented to develop engineered productivity standards for the remaining crews in order to improve their productivity as well focused on six crews. Authors described the application of an industrial engineering philosophy of work measurement of lean production theory and the technique of simulation analysis to capture current work methods, generate and test alternative methods, and develop new productivity standards for drainage maintenance operations crews. Salem et al. [29] proposed a new lean assessment tool to quantify the results of lean implementations. The assessment tool evaluates six lean construction elements including: (1) Last Planner; (2) increased visualization; (3) huddle meetings; (4) first-run studies; (5) five S's; and (6) fail safe for quality. Mao and Zhang [30] developed a construction process reengineering framework and corresponding methodologies that integrate lean principles and computer simulation techniques. Computer simulation techniques are incorporated into the framework to virtually simulate and assess the efficiency and effectiveness of the reengineered construction process that is achieved based on lean principles. Simulation made it easier to quantify and assess the effectiveness and efficiency of the reengineered construction process. Integrating lean principles and computer simulation techniques, the proposed construction reengineering framework was useful and workable in streamlining the construction process for improved productivity, efficiency, and cost effectiveness, which was confirmed by the case study of a tunnel project. Song and Liang [31] observed waste in both project-level contractor coordination and operation level construction performance. A vertically-integrated scheduling system that features location based look-ahead scheduling and graphic weekly work planning was developed to improve project-level contractor coordination. To implement waste elimination solutions at the operation level, construction simulation and 3-D visualization were applied to facilitate lean implementation. Meanwhile, the impact of lean on sustainability was observed and discussed. Deshpande et al. [32] presented the techniques used by a midsize industrial construction contractor: (1) Purpose built facility; (2) Making decisions at the last responsible moment; (3) Lean audits; and (4) 5S in the design of industrial projects to encapsulate and implement techniques of lean production in the management of design. Shewchuk and Guo [33] proposed a lean approach to panel stacking, panel sequencing, and stack locating, where panels within each stack form a continuous structure and are erected via continuous flow. The objectives were to minimize the quantity of stacks, panel material handling distance, and the work required to position and brace panels. The proposed approach and algorithm result in improved performance have no shape restrictions and always provide feasible solutions. Additionally, computational experiments show that the algorithm outperforms methods being employed in the construction industry today.

RESEARCH OBJECTIVES

As a response to the construction problems previously discussed, the research seeks to confirm the following objectives: (1) Determine the implementation of lean ideal; (2) Identify the source of wastes classified under lean construction industry; (3) Examine general perceptions of the construction industry with the lean construction principles of practices; (4) Study reducing and eliminating wastes as classified under development of Last Planner System as a technique of lean construction implementation and to evaluate the effectiveness of implementing last planner to increase plan reliability; (5) Examine the relationship between lean construction and performance improvement programs in construction organizations; and (6) Analyze the characteristics of successful performance improvement programs, and develop a model that identify three critical elements: (a) Time spent on improvement, (b) Improvement skills and mechanisms, and (c) Improvement perspective and goals.

LEAN JOURNEY

Lean implementation begins with leadership commitment and is sustained with a culture of continuous improvement. When the principles are applied properly, dramatic improvements in safety, quality, and efficiency can be achieved at the project level. Improvements at the process and enterprise levels are enablers that make improvements at the project level more successful and allow such improvements to be sustainable [34].

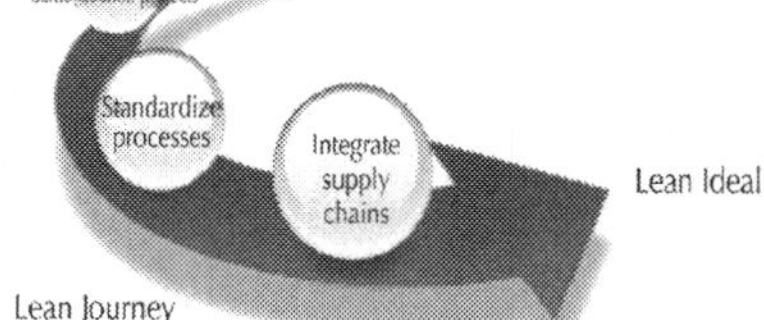

Figure 2: Lean journey to implement lean ideal [34].

The lean ideal is to provide a custom product exactly fit for purpose and delivered instantly with no waste to the subsequent actions that may be necessary in order for projects to pursue that ideal [34]. The ability of individuals and organizations to follow this process will vary with position and circumstances, but to the extent possible, the following should be implemented on projects: (1) Select suppliers who are willing to adopt lean project delivery; (2) Structure the project organization to allow money to move in pursuit of the best project-level returns; (3) Define and align project scope, budget, and schedule; (4) Explore adaptation and development of methods; (5) Make design decisions, with explicit alternatives against stated criteria; (5) Practice production control in accordance with lean principles; (6) Build quality and safety into projects; (7) Implement JIT and multi-organizational processes after site demand; (8) Use evaluations and planning on process that transform materials; (9) Use computer modeling to integrate product and process design; (10) Use 5S workshops: a tool for workplace organization and promoting teamwork (S_1) **S**ort through items, keep what is needed and dispose of what is not; (S_2) **S**traighten: organize and label everything; (S_3) **S**hine: clean; which can also expose abnormal and pre-failure conditions; (S_4) **S**tandardize: develop rules to maintain the first three S's; and (S_5) **S**ustain: manage to maintain a stabilized workplace and initiate continuous improvement when needed and (11) Apply Value Stream Mapping to make visible all the steps in process. These can be organized specially for projects and preceded by a pre-project phase [34].

CONSTRUCTION WASTES

Construction management suffers many problems and the majority is practical which need to be solved or better understood. As a result, the construction industry is overwhelmed by delay and often has suffered cost and time overrun. Alsehaimi and Koskela [35] reported that the poor project management was a dominant and common reason for delay in construction projects. Consequently, these problems associated with management, in particular, should be understood, and efforts need to be directed toward developing solutions and more efficient methods of operation [36]. The introduction of new production philosophies in construction requires new measures of performance Koskela [3], such as waste, value, cycle time or variability. UK studies indicated that up to 30% of construction is rework, only 40–60% of potential labor efficiency, accidents can account for 3–6% of total costs, and at least 10% of materials are wasted. The cost of rework in Australian construction projects has been reported as being up to 35% of total project costs and contributes as much as 50% of a project's total overrun costs. In fact, rework is one of the primary factors contributing to the Australian construction industry's poor performance and productivity [37]. In general, a very high level of wastes/non-value added activities is assumed to exist in construction, and it is difficult to measure all waste in construction. Several partial studies from various countries have confirmed that wastes in construction industry represent a relatively large percentage of production cost. The existences of significant number of wastes in the construction have depleted overall performance and productivity of the industry, and certain serious measures have to be taken to rectify the current situation. Waste measures are more effective to support process management, since they enable some operational costs to be properly modeled and generate information that is usually meaningful for the employees, creating conditions to implement decentralized control. Fig. 3 shows the waste percentages of time in manufacturing and construction.

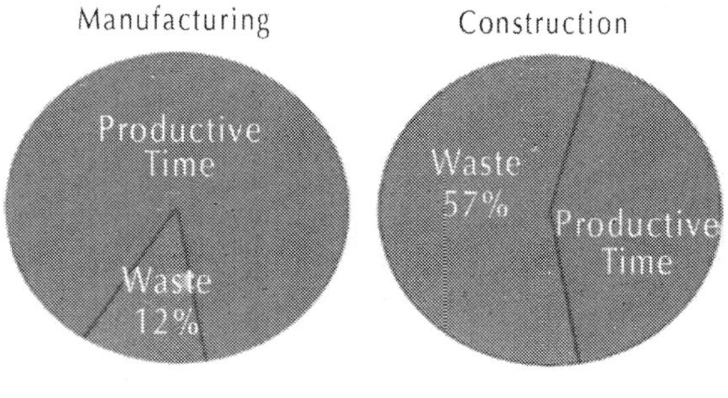

Figure 3: Waste percentages of time in manufacturing and construction [38].

Waste has been defined by Alarcon [39] as "Anything different from the absolute minimum amount of resources of materials, equipment and manpower, necessary to add value to the product." In general, any losses produced by activities that generate direct or indirect costs but do not add value to the product from the point of view of the client can be called "waste." Waste is measured in terms of costs; other types of waste are related to the efficiency of the processes, equipment or personnel, and are more difficult to be measured because the optimal efficiency is not always known. Value adding and non-value adding activities can be defined as follows [39]:

Value adding activities: Those which convert materials and/or information in the search to meet client's requirements and

Non-value adding activities (waste): Those which are time, resource, or space consuming, but do not add value to the product. Waste in the construction industry has been the subject of several research projects around the world in recent years. However, most studies tend to focus on the waste of materials, which is only one of the resources involved in the construction process. This seems to be related to the fact that most studies are based on the conversion model, in which material losses are considered to be synonymous of waste. Formoso et al. [40] stated that many people in the industry have considered wastes are directly associated with the debris removed from the site and disposed of in landfills, and they suggested that the main reason for this relatively narrow view of waste is perhaps the fact that it is relatively easy to see and measure. The main focus for those conventional material waste studies in construction is seen to be restricted to physical waste or material waste in construction and/or the specific impacts due to the physical waste itself.

Formoso et al. [41] had significantly grouped some researches and studies, done by other researchers around the world on the wastes in construction, into two main aspects based on the impacts of the construction waste:

Researches and studies mostly focused on the environmental impacts that result from the generation of material waste, aimed to reduce the generation of waste at source and to propose alternative methods for treatment of construction waste in order to reduce the demand for final disposal areas, others concerned with the measurement and prevention of construction waste, regarding sustainability requirements stated by Dutch environmental policies;

Researches and studies mostly concerned with the economic impacts of waste in the construction industry and concluded that there was a considerable amount of waste that can be avoided by adopting relatively simple prevention procedures. Other researches also pointed out that storage and handling were major causes of waste, while most of the problems concerning waste on building sites are related to flaws in the management system and have very little to do with the lack of qualification of workers. Basically, Koskela[3] has been looking for the evidences of waste and value loss due to (1) Quality of works; (2) Constructability; (3) Material management; (4) Non-productive time; and (5) Safety issues. Formoso et al.[41] proposed the main classification of waste based on the analysis of some Brazilian building sites they had carried out as:

1. *Overproduction*: related to the production of a quantity greater than required or earlier than necessary. This may cause waste of materials, man hours, or equipment usage. It usually produces inventories of unfinished products or even their total loss, in the case of materials that can deteriorate. An example of this kind of waste is the overproduction of mortar that cannot be used on time;
2. *Substitution*: is monetary waste caused by the substitution of a material by a more expensive one (with unnecessary better performance); the execution of simple tasks by an overqualified worker; or the use of highly sophisticated equipment where a much simpler one would be enough;
3. *Waiting time*: related to the idle time caused by lack of synchronization and leveling of material flows and pace of work by different groups or equipment. One example is the idle time caused by the lack of material or by lack of work place available for a gang;
4. *Transportation*: concerned with the internal movement of materials on site. Excessive handling, the use of inadequate equipment or bad conditions of pathways, can cause this kind of waste. It is usually related to poor layout and the lack of planning of material flows. Its main consequences are as follows: waste of man hours, waste of energy, waste of space on site, and the possibility of material waste during transportation;
5. *Processing*: related to the nature of the processing (conversion) activity, which could only be avoided by changing construction technology. For instance, a percentage of mortar is usually wasted when a ceiling is being plastered;
6. *Inventories*: related to excessive or unnecessary inventories which lead to material waste (by deterioration, losses due to inadequate

stock conditions on site, robbery, and vandalism) and monetary losses due to the capital that is tied up. It might be a result of lack of resource planning or uncertainty on the estimation of quantities;

7. *Movement*: concerned with unnecessary or inefficient movements made by workers during their job. This might be caused by inadequate equipment, ineffective work methods, or poor arrangement of the working place;
8. *Production of defective products*: it occurs when the final or intermediate product does not fit the quality specifications. This may lead to rework or incorporation of unnecessary materials to the building (indirect waste), such as the excessive thickness of plastering. It can be caused by a wide range of reasons: poor design and specification, lack of planning and control, poor qualification of the team work, lack of integration between design and production, etc.; and
9. *Others*: waste of any other nature than the previous ones, such as burglary, vandalism, inclement weather, and accidents.

Alarcon [39] divided the controllable wastes into three different activities, which associate with flows, conversions, and management activities: (1) *Controllable causes associated with flows*: (a) Resources: (i) Materials: lack of materials at the work place; materials are not well distributed; inadequate transportation means; (ii) Equipment: non-availability; inefficient utilization; inadequate equipment for work needs; and (iii) Labor: personal attitudes of workers; rebellion of workers; and (b) Information: (i) Lack of information; (ii) Poor information quality; and (iii) Timing of delivery is inadequate; (2) *Controllable causes associated with conversions*: (a) Method: (i) Deficient design of work crews; (ii) Inadequate procedures; and (iii) Inadequate support to work activities; (b) Planning (i) Lack of work space; (ii) Too much people working in reduced space; and (iii) Poor work conditions; and (c) Quality: (i) Poor execution of work; and (ii) Damages to work already finished; and (3) *Controllable causes associated with management activities*: (a) Decision making: (i) Poor allocation of work to labor; and (ii) Poor distribution of personnel; and (b) Ineffective supervision/control: Poor or lack of supervision. Modeling, evaluation of wastes, and performance in construction projects have been a challenge for the construction industry for decades. Several models and procedures have been proposed for the evaluation of project performance at site and project level. Some of these models focus on prediction of project performance while others focus on measuring. Traditional models offer only a limited set of measures as most of them limit their analysis to a number of measures such as cost, schedule, or productivity (usually labor productivity). The shortcomings of the traditional control systems and models are unable or not appropriate to measure those new performance elements but Alarcon [39] suggested that some of the concepts developed in previous researches can be utilized in modeling new performance elements for construction required for continued improvement. It is worthwhile to point out some opinions of different researchers and authors related to the extent of performance elements in the aspects of construction process. He has characterized performance in seven criteria or elements on which management should focus its efforts on as:

- (1) *Effectiveness*: A measure of accomplishment of things;
- *Efficiency*: A measure of utilization of resources. It can be represented as a ratio of resource expected to be consumed divided by the resources actually consumed;
- *Quality*: A measure of conformance with specifications;
- *Productivity*: Theoretically, this is defined as a ratio between output and input, and it is primary measured in terms of cost. In the context of the construction industry, the output is the structure or facility that is built or some components of it. The major input into construction process includes work force, materials, equipment, management, energy, and capital;
- *Quality of work life*: A measure of employee's affective response to working and living in organizational systems. Often, the management focus is on insuring that employees are satisfied, safe, and secure and so forth;
- *Innovation*: This is the creative adaptation process of product, service, process, or structure in response to internal; as well as external; pressures, demands, and changes, needs and so forth; and
- *Profitability*: A measure or a set of measures of relationships between financial resources and uses for those financial resources.

Lean Thinking Principles

There are five fundamental principles for lean thinking, which have to be followed step by step to gain the maximum benefit of the lean success: (1) *Specify Value*: Specify value from customer's own definition and needs and identify the value of activities, which generate value to the end product; (2) *Identify the Value Stream*: Identify the value stream by elimination of everything, which does not generate value to the end product. This means, stop the production when something is going wrong and change it immediately. Processes which have to be avoided are miss production, overproduction (repeat production of the same type of product, etc.), storage of materials and unnecessary processes, transport of materials, movement of labor workforces and products, and finally production of products which does not live up to the wished standard of the customer as well as all kind of unnecessary waiting time; (3) *Flow*: Ensure that there is a continuous flow in the process and value chain by focusing on the entire supply chain. Focus has to be on the process and not at the end product. However, the flow will never get optimal until customer value is specified, and the value stream is identified; (4) *Pull*: Use pull in the production and construction process instead of push. This means produce exactly what the customer wants at the time the customer needs it and always prepared for changes made by customer. The idea is to reduce unnecessary production and to use the management tool "Just In Time"; and (5) *Perfection*: Aims at the perfect solution and continuous improvements. Deliver a product which lives up to customer's needs and expectations within the agreed time schedule and in a perfect condition without mistakes and defects. The only way to do so is by having a close communication with the customer/client as well as managers, and employees are between. Fig. 4summarizes examples of lean tools already used in job sites.

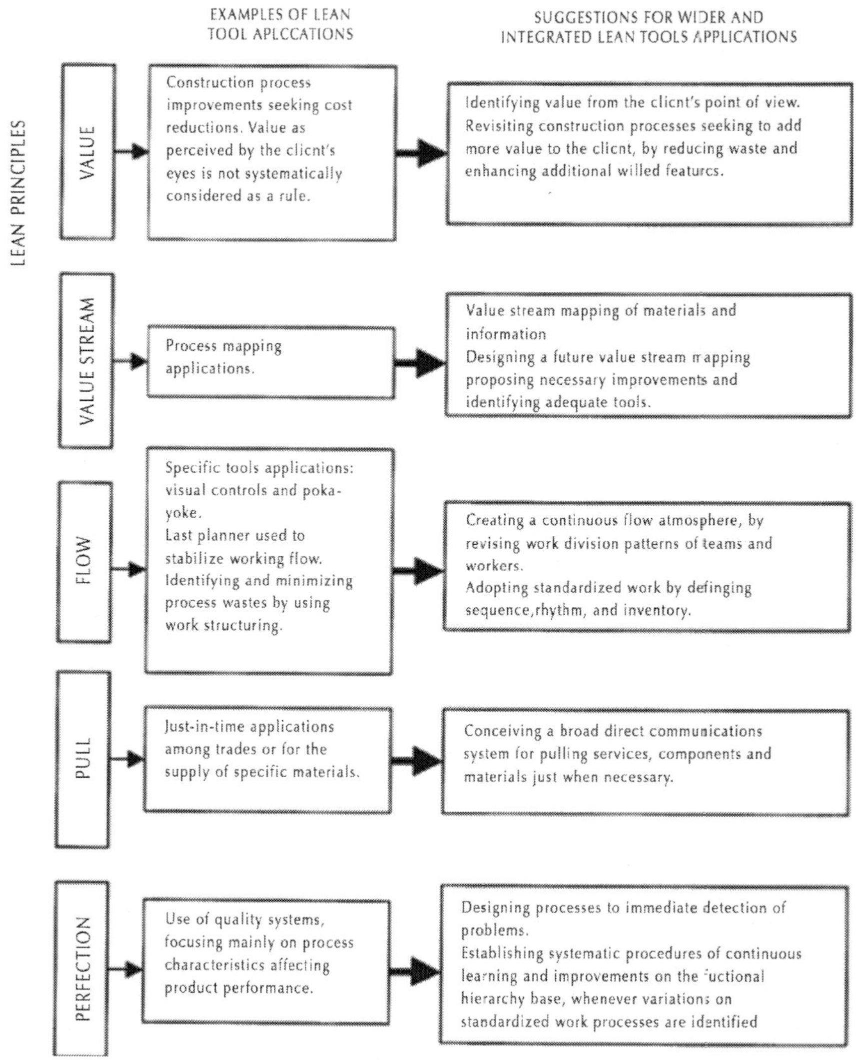

Figure 4: Examples of lean tools in construction implementation and suggestions [42].

Koskela [3] has summarized lean thinking into eleven principles which are (1) Reduce the share of non-value adding activities (waste); (2) Increase output value through systematic consideration of customer requirements; (3) Reduce variability; (4) Reduce cycle times; (5) Simplify by minimizing the number of steps, parts and linkages; (6) Increase output flexibility; (7) Increase process transparency; (8) Focus control on the complete process; (9) Build continuous improvement into the process; (10) Balance flow improvement with conversion improvement; and (11) Benchmark. There are fourteen principles organized in four categories: (1) Philosophy; (2) Process; (3) People and Partners; and (4) Problem Solving, as seen in Fig. 5[43].

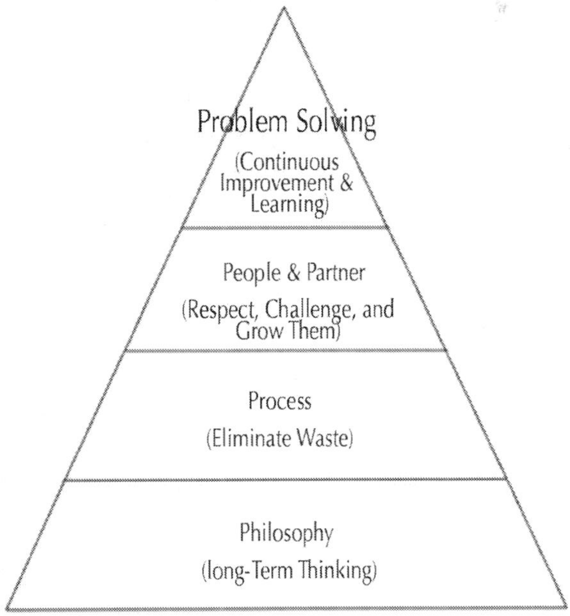

Figure 5: "4P" of the lean way [43].

The fourteen (14) management principles of the lean way are as follows: (1) Base decisions on long-term philosophy even at the expense of short-term financial goals (Philosophy); (2) Create continuous process flow to bring problems to the surface (Process); (3) Use "Pull" systems to avoid overproduction (Process); (4) Level out the workload (Process); (5) Build a culture of stopping to fix problems to get quality right the first time (Process); (6) Standardized tasks are the foundation for continuous improvement and employee empowerment (Process); (7) Use visual control so no problems are hidden (Process); (8) Use only reliable, thoroughly tested technology that serves people and processes (Process); (9) Grow leaders who thoroughly understand the work, live the philosophy, and teach it to others (People and Partners); (10) Develop exceptional people and teams who follow your company's philosophy (People and Partners); (11) Respect your extended network of partners and suppliers by challenging them and helping them improve (People and Partners); (12) Go and see for yourself to thoroughly understand the situation (Problem Solving); (13) Make decisions slowly by consensus, thoroughly considering all options; implement rapidly (Problem Solving); and (14) Become a learning organization through relentless reflection and continuous improvement (Problem Solving).

LEAN CONSTRUCTION TECHNIQUES

Lean construction is a way to design production systems to minimize waste of materials, time, and effort in order to generate the maximum possible amount of value [4]. Lean Construction is using the same principles as lean production to reduce waste and increase the productivity and effectiveness in construction work. The most important determinants of construction are supposed to be workflow reliability and labor flow, but lean construction has changed the traditional view of the project as transformation, and embraces the concept of flow and value generation. Similarly, it shares the same objectives of lean production, e.g., cycle time reduction, elimination of waste, and variability reduction. Continuous improvement, pull production control, and continuous flow have been the direction for the implementation of lean construction. Lean construction is composed of the following techniques [44]: (1) *Concurrent Engineering*: Concurrent engineering can be described as parallel execution of various tasks by multidisciplinary teams with the goal of obtaining most favorable products concerning functionality, quality, and productivity. Many enhancements can be accomplished by using concurrent engineering. Scheduling could be recovered by network analysis (CPM and PERT). Many other opportunities can be achieved through overlapping activities, splitting activities and reducing the transfer time between different activities. The important planning parameters for scheduling concurrent activities are lead time, quantity, and risk under ambiguity. Concurrent engineering is focusing on the team efforts; communication and information sharing are the keys for discovering new ideas. While partnering with subcontractors and suppliers can also be good changes regarding concurrent engineering, the success of lean production is depending on the involvement of all participants in the early stages of the design; (2) *Last planner*: The last planner is the person or group of people responsible for production unit control, which means completion of individual tasks at the operational level. Last planner necessitates work flow control, ascertaining the stream of supply, design, and installation throughout production units. This can only be done by using look-ahead schedule, which determines the progression and rate of work. It carves up the master schedule into many packages, specifying the techniques of check capacity, execution, and establishes a stockpile of standing by work. The scope of look-ahead schedule ranges from 2 to 6 weeks and should be put in order by team work; (3) *Daily huddle meetings*: Daily huddle meetings provides a platform for the team members to share their views and to share what has been achieved, at the same time, discus problems they are facing during the production process; (4) *The Kanban System*: The strategy of Kanban is grounded on key components, i.e., market place, supplier kanbans, collection vehicle, satellite stores, and inventory management system. Market places are site warehouse that allocate different materials and small tools to the workers. Similarly, satellite stores are situated on site, where they get products from market places. Collection vehicle collects materials from preferred suppliers to the operational site. Kanban use plastic bins as a signal to pull materials from suppliers to site, using the concept of Just In Time. Request forms are normally used as kanban signals between market place and satellite stores. The system of kanban starts normally with open doors, so that the site can pull materials from the supplier up to certain perimeters. Subsequently, the material requested from suppliers arrives at market, and products are later on picked from the stores, which are usually managed by recorder points; (5) *Plan Conditions and Work Environment in the Construction Industry (PCMAT)*: The purpose is to introduce a plan of health and safety into the project execution, called "Plan of Condition and Work Environment." These safety activities can generate limitations for scheduled tasks and that is why it should be embraced as a part of assignments. All safety practices are therefore amalgamated in short-term planning, which can be analyzed through daily feedback from crew and subcontractors respectively; (6) *Quality Management Tools*: The fusion of quality management tools in the lean construction is based on the change from conformance based quality to the quality at the source. A point system is normally employed to evaluate the execution of planned controls, which will help workers to follow planned controls instead of quality corrections; and (7)*Visual Inspection*: Visual inspection shows the uneven nature of the construction and leads to the application of visual tools for material, work and information flow, etc. Identification of materials can accelerate repetitive processes and diminishes the risk of selecting wrong product. Progress charts and schedules can implement the dedication to the completion of tasks. Information and technology can also improve the communication between decision maker and executer, and can accelerate the process as well. The Lean Construction Institute [38] described how current projects are to be managed and defines the project management as follows: (1) Determine client requirements and design to meet them; (2) Align design to quality, schedule, and budget limits; (3) Manage the project by breaking it into pieces, estimating duration and resource requirements for each piece, and then put the pieces in a logical order with Critical Path Method (CPM); (4) Assign or contract for each piece, give start notice and monitor each piece to assure it meets safety, quality, schedule and cost standards. Take action on negative variance from standards; (5) Coordinate using the master schedule and weekly meetings; (6) Cost may be reduced by productivity improvement; (7) Duration may be reduced by speeding each piece or changing logic; and (8) Quality and safety get better with inspection and enforcement.

APPLICATION CHANNELS OF LEAN CONSTRUCTION

Lean Project Delivery System (LPDS)

The Lean Project Delivery System is a set of interdependent functions, rules of decision making, procedures for execution of functions, and as implementation aids and tools, including software when appropriate, and is a conceptual framework developed by Ballard [45] to guide the implementation of lean construction on project-based production systems. LPDS was depicted as a model with five main phases, where each phase is comprised of three modules; see Fig. 6. The interdependence between the phases was represented by sharing one module between two subsequent phases. Production control and lean work structuring were both shown to extend throughout the five main phases. Learning was introduced to underscore the need to document lessons learned from one engagement to another.

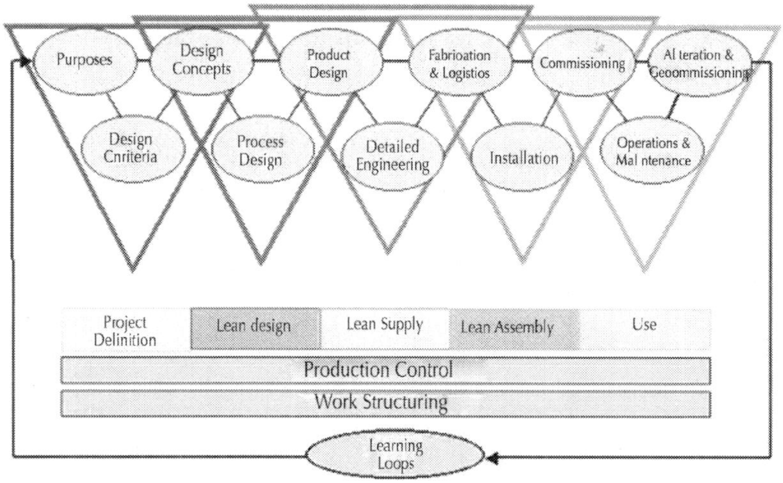

Figure 6: Lean project delivery system [43].

The LPDS model consists of 15 modules, 11 organized in 4 interconnecting triads extending from project definition to design to supply and assembly, plus 2 production control modules and work structuring module, both conceived to extend through all project phases, and the post-occupancy evaluation module, which links the end of one project phase to the beginning of the next: (1) *Project Definition*: (a) Needs and Values Determination; (b) Design Criteria; and (c) Conceptual Design; (2) *Lean Design*: (a) Conceptual Design; (b) Process Design; and (c) Product Design; (3) *Lean Supply*: (a) Product Design; (b) Detailed Engineering; and (c) Fabrication/Logistics; (4) *Lean Assembly*: (a) Fabrication/Logistics; (b) Site Installation; and (c) Testing/Turnover; and (5) *Production Control*: (a) Work Flow Control; and (b) Production Unit Control. Work Structuring and Post-Occupancy Evaluation are thus far only single modules. Essential features of LPDS are as follows [45]: (1) Structure and management of a project are aimed at creating value; (2) Cross-functional teams, involved in front end planning and design, include members from all areas of production process; (3) Project control would be a tool executed throughout the project as opposed to reliance on after the fact variance detection; (4) Optimization efforts are focused on making work flow reliable and not to focus on improving productivity; (5) Pull techniques are used to govern the flow of materials and information; (6) Capacity and inventory buffers are used to absorb variability in the production process; (7) Feedback loops incorporated at every level, are aimed at a rapid system adjustment and learning; and (8) Work structuring of the entire process increases value and reduces waste at the project delivery level. Efforts to improve performance at the planning level increases performance at project level.

Last Planner System (LPS)

One of the most effective ways to increase efficiency of construction industry is to improve planning and control process. In Lean Construction, planning and control are considered to be complementary and dynamic processes maintained during the course of the project. Planning defines the criteria and creates strategies required to reach project objectives, control makes sure that each event will occur following the planned sequence. Re-planning must be done when the previously established sequences are no longer applicable or convenient. Feedback facilitates learning when the events do not occur as planned [15] and [45]. One of the best known Lean techniques is the Last Planner System which has been demonstrated to be a very useful tool for the management of construction process, and continuous monitoring of the planning efficiency, to assist in developing foresight, smoothing workflow variations, and reducing/removing uncertainties plaguing construction processes. It consists of work flow control and production unit control. Work flow control is accomplished primarily through the look-ahead process, while production unit control is accomplished primarily through weekly work planning. Mossman [46] defined the last planner as a system for collaboratively managing the network of relationship and conversations required for program coordination, production planning and project delivery, by promoting conversations between trade foreman and site management at appropriate levels of detail before issue become critical [46]. Last Planner System aims to shift the focus of control from the workers to the flow of work that links them together. The two main objectives of LPS are to make better assignments to direct workers through continuous learning and corrective action and to cause the work to flow across production units in the best achievable sequence and rate. The last planner integrated components are: master plan, phase planning, look-ahead planning, weekly work planning, Percentage of Promises Completed on time or Percent of Planned Completed "PPC" (A measure key of the Last Planner System success) and reasons for incompleteness, when systematically implemented can bring many advantages and add major benefits to construction management practice in general and planning practice in particular. PPC does not measure productivity or production, only planning effectiveness. But of course, the values of PPC are related to production and productivity indirectly, it is assumed that when a project team improves its planning it reduces variation, and thus can become more productive by matching its production resources more closely to the demand for them, so reducing waste. When the weekly work plan is executed, an analysis of the previous week report is made, and the PPC is calculated by dividing the quantity of works effectively completed by the total quantity of works that had been planned. A note explaining the reasons justifies any work that had been planned but was not completed. When the PPC is calculated, a re-programming of the services is made, indicating the services that had already been executed and those that had been planned but were not executed. The immediate result of this re-programming is the calculation of a new date for finishing the construction [47]. Companies utilizing the LPS have been able to maintain project on time and at budget, as well as having a stress-free production planning and control process. Fig. 7 illustrates the results of implementation of the LPS on a construction project, which clearly reflects the positive impact of the system on budget and productivity [48]. Fernandez et al. [49] reported benefits attributed to LPS implementation were as follows: (1) smooth work flow, (2) predictable work plans, (3) reduced cost, (4) reduced time of project delivery, (5) improved productivity, and (6) greater collaboration with field personnel and subcontractors. Test case projects also reported certain challenges faced by project participants when applying LPS: (1) lack of leadership, (2) organizational inertia, (3) resistance to change, (4) lack of training, (5) contractual issues, and (6) lack of experience and knowledge, among others. Last Planner System (LPS) has four main elements [46]: (1) *Programming Workshop*: Collaboratively creating and agreeing production sequence (and compressing it if required); (2) *Make-Ready*: Making tasks ready so that they can be done when we want to do them; (3) *Production Planning*: Collaboratively agreeing production tasks for the next day or week; and (4) *Continual Improvement*: Learning about and improving the project, planning and production processes.

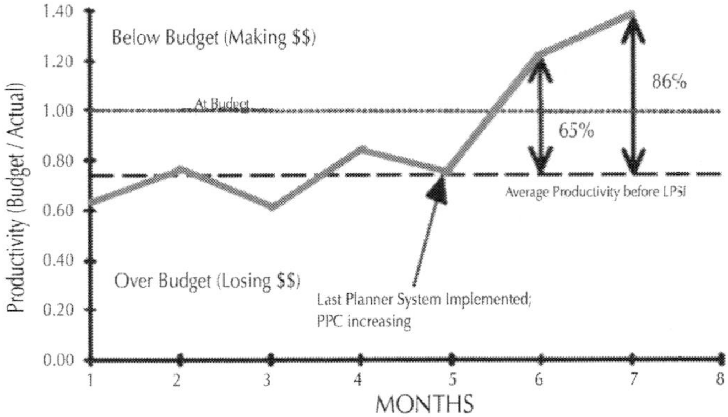

Figure 7: Productivity improvement using LPS [48].

Table 1 provides a comparison of the differences between CPM and last planner, identified CPM as Strategic Planning and last planner as production planning.

Table 1: Separate strategic planning from production planning

Critical path method	Last Planner System
•CPM logic embedded in software	•Applied common sense
•High maintenance	•Low maintenance
•Managing critical path	•Managing variability
•Focus on managing work dates	•Focus on managing work flow
•Planning based on contracts	•Planning based on interdependencies

The Last Planner System is based on a traditional planning system, and is performed by field foremen; the condition of (SHOULD-CAN-WILL-DID) is implemented. Figure 8.1 and Figure 8.2 represented the diagram of the Last Planner System.

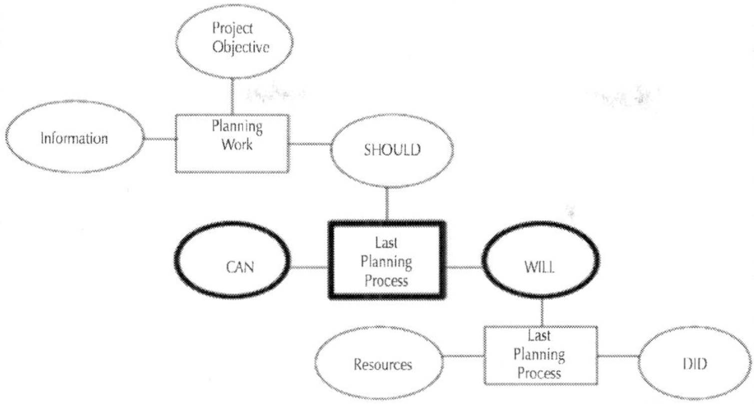

Figure 8.1: Last planner system [15].

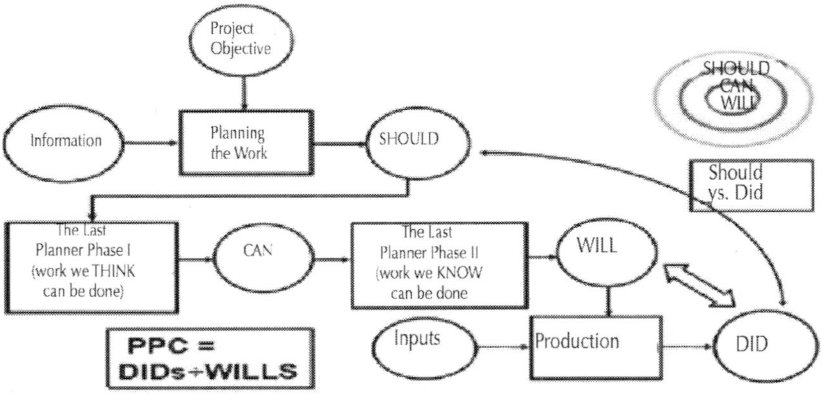

Figure 8.2: Last Planner System [38].

A reliable assignment, one that gets done at the required time, determines what WILL be done, after considering both what SHOULD from higher-level schedules and what CAN be done based on the situation at hand. Assignments are likely to get done when they are well defined, resource sound, in the right sequence, and within the capacity of the crew. The last planner's job is to make certain task in the assignment that meets these criteria and to reject assignments that do not. Last planners can reasonably commit to completing the tasks on weekly work plans that meet these criteria. To be effective, production management systems must tell what should be done, what can be done, and what will be done; then, they compare what was done to improve planning [15]. The term SHOULD is considered as: Hopefully; CAN means: Probably; and WILL means: Absolutely. Fig. 9 illustrates the possible relationships among SHOULD, CAN, and WILL.

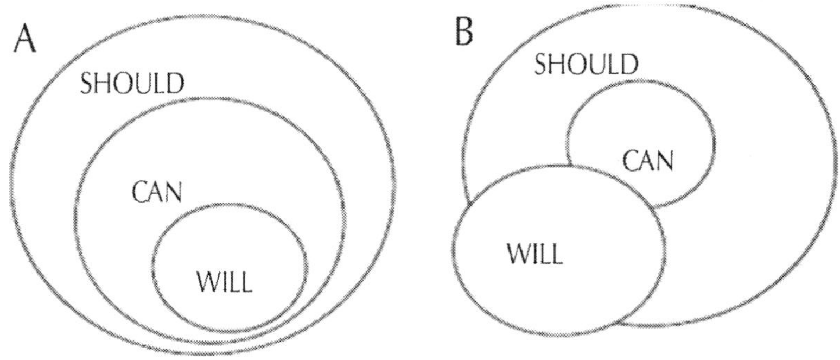

Figure 9: Diagram of SHOULD-CAN-WILL [15].

Diagram (A) in Fig. 9 represents a scenario with the highest probability of task completion, and diagram (B) shows the certainty of failure. Referring to Fig. 8.1, a reliable assignment determines what WILL be done, after considering what SHOULD and CAN get done based on the situation at hand. The assignments in diagram (A) are well defined, sound, in the right sequence, and doable by the crews. Thus, the task is likely to get done at the required time. In contrast, the assignments in diagram (B) are out of plan and have much variability to be controlled. Thus, the probability of task completion decreases. The Last Planner System has four levels [50]: (1) *Master Schedule*: setting milestones and strategy identification of long lead items; (2) *Phase Schedule*: Pull planning (specifies handoffs; identify operational conflicts); (3) *Look-ahead Plan*: Make Work Ready Planning (to ensure that work is made ready for installation; re-planning as necessary); and (4) *Weekly Work Plan (WWP)*: commitments to perform work in a certain manner and a certain sequence, and Learning (measuring percent of planned completed PPC), deep dive into reasons for failure, developing and implementing lessons learned. Fig. 10 describes the detailed processes of LPS.

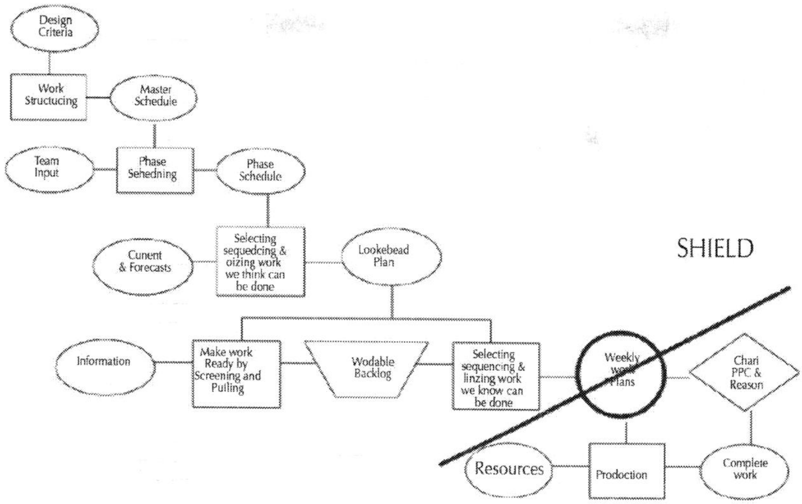

Figure 10: Process of last planner system [50].

Fig. 11 illustrates the sequence of implementing the Last Planner System, and Fig. 12 illustrates how Last Planner System achieves lean concept.

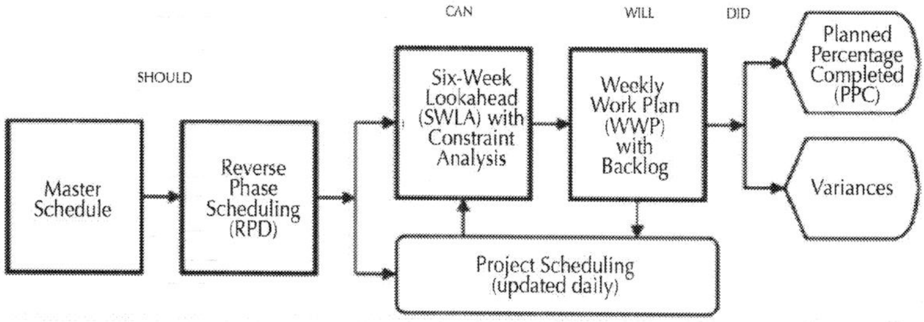

Figure 11: The sequence of last planner process.

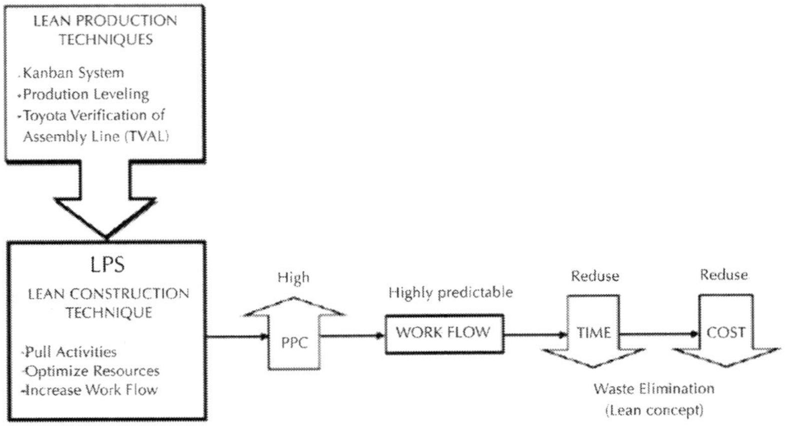

Figure 12: How Last Planner System achieves lean concept.

Wambeke et al. [51] provided quantitative data that demonstrated how using the LPS method reduced and/or eliminated variation for the mechanical contractor involved in construction project. Also they used a risk assessment matrix as a new and effective means of prioritizing which causes of variation should be targeted first for reduction. Kim and Ballard [52] investigated the theories implicit in two prevalent project control systems: (1) Earned Value Method (EVM) and (2) Last Planner System (LPS). They introduced two fundamental and competing conceptualizations of management: (1) Managing By Means (MBM) and (2) Managing By Results (MBR). The EVM is found to be based on MBR. However, project control based on MBR is argued to be inappropriate for managing at the operational level where tasks are highly interdependent. The LPS is found to be based on the MBM view. The empirical evidence from case study suggested that the MBM view is more appropriate to manage works when it is applied to the operation level where each task is highly interdependent.

PERFORMANCE IMPROVEMENT PROCESS MODEL

Fig. 13 shows the key factors that determine the success of improvement process. The model is depicted as a causal loop diagram and illustrates the interactions between key factors.

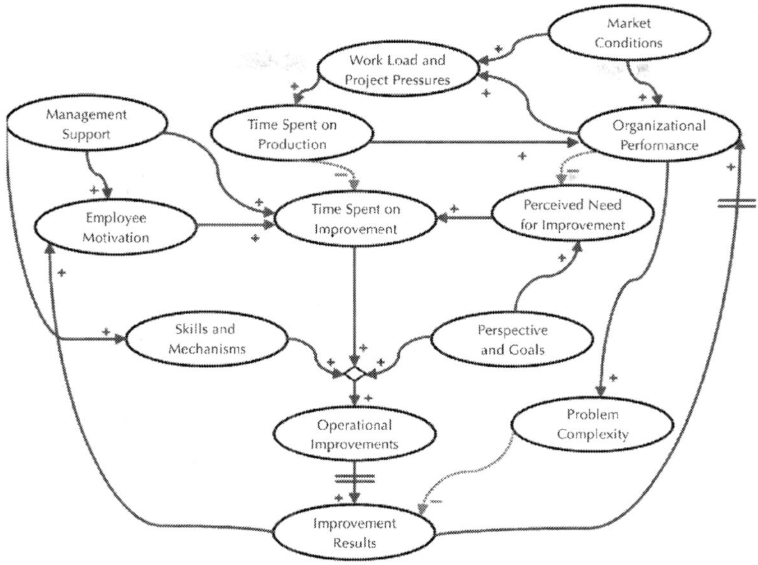

Figure 13: Model of performance improvement process.

An arrow between factors means that factor X affects factor Y. A positive sign indicates that if factor X increases, then factor Y also increases. A negative sign indicates that if factor X increases, then factor Y decreases. A double line indicates a time lag. When more than one arrow converges to a diamond, then ALL of conditions need to be present for the resulting factor to occur and must ALL be present for effective (Operational Improvements) to occur. The development of (Operational Improvements) depends on three key factors: (1) Time spent on improvement; (2) Performance Improvement skills and Mechanisms; and (3) Perspective and Goals. Operational Improvements are the changes implemented by the organization. These improvements result in (Improvement Results) but with a time lag.

Time spent on improvement

Time Spent on Production

(Time Spent on Production) reduces (Time spent on improvement); (Work Load and Project Pressures) increases (Time Spent on Production);

(Market Conditions) increase the (Work Load) because of the increased volume of work, and the difficulty to hire qualified people in a growing market, project staff is spread (thin) and cannot allocate much time to improvement; and (Time Spent on Production) increases the (Organizational Performance). This illustrates the managerial dilemma between (Today Performance) versus (Future Performance).

Management Support

(Management Support) increases (Time spent on improvement). The construction literature considers senior management support critical for the success of improvement effort. (Management Support) is indicated by the following: (a) Personal involvement in improvement efforts; (b) Acknowledging and rewarding the efforts and successes; (c) Hiring employees who can contribute to improvement; (d) Evaluating middle management (project managers and superintendents) based partially on their contributions to improvement efforts; and (e) Providing resources for training and bringing in external experts as needed. Management commitment is also reflected in the approach to risk taking and experimentation. (Management Support) also increases (Employee's Motivation). Employees are less likely to be involved in an improvement effort if their supervisor does not support their involvement. Many supervisors discourage their subordinates to spend time on improvement because it diverts employee attention from real work.

Employee Motivation

(Employee Motivation) increases (Time spent on improvement). In every organization having a typically small percentage of employees who are actively looking for ways to improve work and initiate improvements. These are the champions who put a lot of personal time in improvement. Another group of employees is willing to try new ideas even if they do not make any particular effort to initiate changes. And finally, there is group that is not interested in improvement. (Improvement Results) increase (Employee Motivation) (a) if efforts and successes are acknowledged and rewarded; and (b) if the positive results come fast. If the results take a long time, the participants' motivation is reduced. However, many of the complex production problems may have a longer time lag between the start of improvement effort and the result (the easy/fast solutions typically have small effect and do not bring substantial change). Both management and employees involved in improvement process need to understand this.

Perceived Need for Improvement

(Perceived Need for Improvement) increases (Time spent on improvement). The (Perceived Need for Improvement) is the gap between organizational performance and target performance. Thus, (Organizational Performance) reduces the (Perceived Need for Improvement). Good (Market Conditions) increase (Organizational Performance). In a good market (when the work volume and profit are good) managers perceive less need for improvement even without equally-good operational performance. Furthermore, in a market where the demand is high, the project budgets have higher contingencies, which reduce the pressure for high process performance. Finally, (Improvement Results) increase (Performance) thus reducing the (Perceived Need for Improvement), as they reduce performance pressures. The (Need for Improvement) is directly affected by the (Perspective and Goals) of improvement process. First, improvement goals create a pressure for improvement when the gap between goals and performance widens. Thus, managers can increase the (Perceived Need for Improvement) by setting high performance goals. Benchmarking against (world class) companies is one example. However, it is not only the level of goals, but also the type of goals that generate need for improvement. Even more important, the managerial perspective (and mental models) of the construction process affect the interpretation of the root causes of problems. This issue is discussed later under (Perspective and Goals).

Performance Improvement Skills and Mechanisms

Performance Improvement Mechanisms

The mechanisms for learning can be grouped in three categories: (1) *Learning from Experience*: Such mechanisms include observation and analysis of existing processes (office or field), after action reviews, and any methods for review and evaluation of organizational activities. Work methods improvement developed systematic approaches to analyze production operations; (2) *Gathering Intelligence*: Another way to identify potential improvements is by monitoring the external environment. This includes exploring developments outside the company, keep up with new designs, methods and technologies that take place outside the company; and (3) *Learning through Experimentation*: Experimentation includes the usage of new untested methods and techniques. These could be production technologies, management methods (e.g., last planner), new information systems, incentives systems, etc. Two important issues related to experimentation are: (a) The extent that management supports risk taking (otherwise, no real risks will be taken); and (b) How can we reduce/better control the risk involved in construction experiments? (e.g., it may be necessary to team-up with owners to conduct some experiments). The use of learning mechanisms increases the organization's ability to identify problems and improvements. However, the ability to identify effective changes also depends heavily on the available skills.

Performance Improvement Skills

Every performance improvement process includes three major steps: (1) Acquiring information, when the organization collects measurements, observations, and data (such as statistical data of defects, market data, and productivity data); (2) Interpreting information, when the organization analyzes data to understand what it means, and what are the cause-effect relationships at work, and what are the real causes of observed data; (3) Applying information, when the organization develops and implements improvement initiatives. To perform these steps effectively, the organization needs skills in acquiring relevant and meaningful information, as well as skills in analyzing the information, and creating effective changes. Process Analysis Skills and Root Cause Analysis are essential in order to discover the key factors affecting performance, and develop effective interventions. Without such skills, solutions tend to address symptoms near the problem, rather than root causes. (Management Support (is essential for the development of (Improvement skills and Mechanisms) as they provide the resources for development of Improvement skills through internal training or external experts, and forums for intelligence gathering, after action reviews, and process analysis. Furthermore, experimentation directly depends on the extent that management supports risk taking.

Improvement Perspective and Goals

The term (Perspective) refers to whether the improvement process is Result-focused or Process-focused. The Critical Success Factors are an example of a (Result-focused) approach. Critical Success Factors are those result areas (such as Schedule, Safety, Estimating, Quality, Cost, and Change Management) that directly affect the performance of the organization. TQM and Lean Construction are (Process-focused). The different focus of the improvement process has important implications for the direction of improvement efforts as it leads to the following differences.

Different Goals

Different goals regarding what the improvement teams are trying to accomplish. Result-based goals are typically oriented toward customer expectations. For example, the schedule improvement goal in a result-focused approach is expressed as (Complete all project on or ahead of the promised schedule) versus (Reduce cycle time of process X) from a process-focused perspective. Quality goal may be (Zero Punch list at time of completion) (result-focused) versus (Eliminate Defects and Rework) (Process-focused). Result-focused goals and process-focused goals are both needed, but at different organizational levels. At the strategic level, management needs to establish result-focused strategic improvement goals in the areas that are critical for competitiveness (such as schedule and cost reduction and safety and quality improvement). But in order to meet the strategic improvement goals, the improvement efforts need to focus on the production processes. Traditional management systems do not focus on production processes, but are result-oriented. However, a results attitude emphasizes fixing problems and fighting fires, rather than preventing problems, planning and learning. The key point is that result-focused goals emphasize results with or without process improvement. Such goals have limited effect on (Perceived Need for Improvement) when the results are satisfactory. On the other hand, process-focused goals continue to drive process improvement even if project results are satisfactory. It was suggested that management must focus on process improvement first and results second.

Different Perspectives

Different perspectives regarding the root causes of performance problems. The simple truth is that when there is no explicit focus on the process, the direction of improvement efforts is determined by the prevailing mental models of the participants. The prevailing perspective (mental model) in construction considers project work as a collection of (activities) rather than a flow [3]. People who hold this perspective believe: (1) Sources of the problems are (outside the process) the owner makes changes or adds scope, the design is incomplete, the subcontractors were late, did not provide manpower when needed, etc.; (2) Performance problems are typically attributed to individual factors, such as responsibility, motivation, and skills, rather than systemic factors (how the work is managed, coordinated, etc.); (3) Delivering a project is just like skinning a cat. There are thousand ways to do it, all pretty much the same. Process-focused approaches (such as TQM and lean production) emphasize both the components of the process (activities, crews, etc.) and the interdependencies between components of the complex production system [53]. Sources of the problems are (inside the process) interdependence and variation, and the incentives, behaviors, and work rules that generate and propagate them. Consequently, different perspectives lead to different directions of improvement efforts. In other words, the definition of the problem drives the solutions.

Operational Improvements

Depending on the (Time spent on improvement), (Skills and Mechanisms), and (Perspective and Goals), the amount and type of operational improvements vary. Result-focused programs are more likely to focus on responsibility and accountability, skills and motivation. Applied in a general contractor organization, these programs lead to greater emphasis on contractual clauses (allocate responsibility), pushing the contractors harder (hold them accountable), training project personnel in identification of defects, or increasing efforts (e.g., having more inspections earlier). Similarly, subcontractors may focus on workers' skills, and efficient equipment. This perspective usually does not aim to change the way by which the work itself is done; rather it changes the context within which work is done. On the contrary, process-focused efforts emphasize the interdependencies between process participants, requirements and the work processes themselves. This leads to very different solutions. Result-based improvement efforts may even increase the (waste) in the process (e.g., by adding inspections, and increasing tracking of defects, rather than reducing waste by preventing defects). Finally, the different perspectives also lead to different participation in the improvement process. Result-focus efforts do not lead to continuing cross-functional or cross-organizational efforts because they do not emphasize the interdependencies between process participants. Cross-organizational cooperation is typically limited to project-level initiatives, but there is no long-term cooperation between contractors, designers and owners to continuously improve work processes.

Problem Complexity

(Problem Complexity) reduces (Improvement Result) assuming that the perspective and skills remain the same. The problems that the improvement effort addresses, have different levels of complexity. Simple problems involve few organizational units/functions and have a simple methodology. An example of a simple system is a crew that performs a relatively simple operation under conditions of low variability, e.g., painting. There is only one organizational unit involved (the painting crew) and the operation has few steps. The factors affecting the performance are relatively few and easily identified (e.g., crew skills, tools and equipment, etc.) The impact of such changes is immediate. The complexity of problems increases as the number of organizational units and their interactions increase, and as the number of steps required and their variability increases. For example, the construction of foundation includes layout, excavation, forming, rebar installation, concrete pour, and stripping the forms. This is a relatively complex operation that involves several different organizational units, and multiple steps and interactions. Improving performance of complex operations requires cross-functional (or even cross-organizational) changes in the way the work is organized and managed (in term of sequence, interdependencies, technologies, incentives, control mechanisms, etc.). Thus, complex problems require process-focus and cross-organizational efforts. As a result, as problem complexity increases, it becomes harder to achieve improvement results.

Improvement Results and Feedback Loops

(Improvement Results) increase (Organizational Performance) although with a time lag (for example, the results of training will be observed in later phases or following projects). (Improvement Results) also increase (Employee Motivation) as well as management support, which leads to spending more time on improvement. This creates a positive feedback loop. On the other hand, when organizational performance increases, the work load typically increases because the organization is more successful in getting more work. In addition, the perceived need for improvement also declines. Increased work load and reduced need for improvement reduce the Time spent on improvement. Thus, a negative feedback loop is created that (regulates) the process. (Organizational Performance) increases (Problem Complexity) that is, as the organization performance increases, further improvements require solutions to more complex problems. Effective improvements are harder to identify and implement, with fewer and slower improvement results. This creates another negative feedback loop. That means if improvements are based on training, motivation, and extra work load (such as additional inspections), the organization will have to increase its efforts simply to maintain the same level of performance. However, when improvements are incorporated in work processes (rather than people or inspections) they can be sustained with fewer efforts.

CONCLUSIONS

This research seeks to confirm the following objectives: (1) Determine the implementation of lean ideal; (2) Identify the source of wastes classified under lean construction industry; (3) Examine general perceptions of the construction industry with the lean construction principles of practices; (4) Study reduction and elimination of wastes as classified under development of Last Planner System as a technique of lean construction implementation and to evaluate the effectiveness of implementing last planner to increase plan reliability; (5) Examine the relationship between lean construction and performance improvement programs in construction organizations; and (6) Analyze the characteristics of successful performance improvement programs, and develop a model that identifies three critical elements: (a) Time spent on improvement, (b) Improvement skills and mechanisms, and (c) Improvement perspective and goals. The authors identify different ways to structure improvement program: outcome focused (such as Critical Success Factors) and process-focused (such as Lean Construction). The paper discusses the implications of the different perspectives and argues that they lead to different improvement approaches each reflecting different paradigms for the nature of the change. The authors propose that result-focused improvement programs may be a barrier to the adoption of Lean Construction. The paper proposed a dynamic model of performance improvement process. The model examined the factors affecting the process and their interactions. The paper proposed that: (1) Direction of the improvement effort is strongly influenced by the structure and goals; and (2) Result-focused programs have limited ability to address complex systemic problems. One question for future research is what drives a contractor to establish a result-focused or a process-focused program. It appears that specialty contractors are more familiar with the process-perspective because of their familiarity with productivity improvement studies (which is a process analysis of a relatively simple problem). On the other hand, general contractors are more likely to emphasize overall project results. Future research also needs to (1) Develop and validate a more complete model of performance improvement; (2) Further examine the behavior of improvement process over time; and (c) Use the model as a starting point for system redesign by adding loops and breaking links.

REFERENCES

1. Construction Industry Institute (CII) Lean Principles in Construction, vol. 191-1The University of Texas at Austin (2005) research summary, pp. 1–44
2. J. Womack, D. Jones, D. Roos The Machine that Changed the World: The Story of Lean Production 1st Harper Perennial Ed., New York (1991)

3. L. Koskela, Application of the New Production Philosophy to Construction, Technical Report No. 72, CIFE, Stanford University, CA, 1992.
4. L. Koskela, P. Huovila, J. Leinonen Design management in building construction: from theory to practice Journal of Construction Research, 3 (1) (2002), pp. 1–16
5. S. Bertelsen, L. Koskela, Managing the three aspects of production in construction, in: Proceedings of the 10th Annual Conference of the International Group for Lean Construction, Gramado, Brazil, 2002.
6. P. Koushki, K. Al-Rashid, N. Kartam Delays and cost increases in the construction of private residential projects in Kuwait Construction Management Economics, 23 (3) (2005), pp. 285–294
7. R. Sacks, M. Goldin Lean management model for construction of high-rise apartment buildings Journal of Construction Engineering and Management, 133 (5) (2007), pp. 374–384
8. H. Guo Rethinking Construction Project Management using the VP-based Manufacturing Management Model The Hong Kong Polytechnic University, Hong Kong (2009)
9. P. Teicholz, Labor Productivity Declines in the Construction Industry: Causes and Remedies, Viewpoints, <http://www.aecbytes.com>, 2004 (Retrieved 14.04.04)
10. H. Li, P. Love, Visualisation of building interior design to reduce rework, in: Proceedings of Second International Conference on Information Visualisation, London, UK, 1998, pp. 187–191.
11. P. Love, P. Mandal, J. Smith, H. Li Modeling the dynamics of design error induced rework in construction projects Construction Management Economics, 18 (5) (2000), pp. 567–574
12. S. Green The missing arguments of lean construction Construction Management and Economics, 17 (2) (1999), pp. 133–137
13. F. Gleeson, J. Townend Lean Construction in the Corporate World of the U.K. Construction Industry University of Manchester, School of Mechanical, Aerospace, Civil and Construction Engineering (2007)
14. L. Koskela Lean production in construction L. Alarcon (Ed.), Lean Construction, Balkema, Rotterdam (1997), pp. 1–9

15. G. Howell, G. Ballard, Implementing lean construction: understanding and action, in: Proceedings Sixth Annual Conference of the International Group for Lean Construction, Guaruja, Sao Paulo, Brazil, 1998.
16. A. Conte, D. Gransberg Lean construction: from theory to practice AACE International Transactions, CS10 (2001), pp. 1–5
17. G. Wright Lean construction boosts productivity Building Design and Construction, 41 (12) (2000), pp. 29–32
18. C. Miller, G. Packham, B. Thomas Harmonization between main contractors and subcontractors: a prerequisite for lean construction? Journal of Construction Research, 3 (1) (2002), pp. 67–82
19. H. Thomas, M. Horman, U. Souza, I. Zavrski Reducing variability to improve performance as a lean construction principle Journal of Construction Engineering and Management, 128 (2) (2002), pp. 144–154
20. H. Thomas, M. Horman, R. Minchin, D. Chen Improving labor flow reliability for better productivity as lean construction principle Journal of Construction Engineering and Management, 129 (3) (2003), pp. 251–261
21. H. Bashford, K. Walsh, A. Sawhney Production system loading cycle time relationship in residential construction Journal of Construction Engineering and Management, 131 (1) (2005), pp. 15–22
22. R. Sacks, A. Esquenazi, M. Goldin LEAPCON: simulation of lean construction of high-rise apartment buildings Journal of Construction Engineering and Management, 133 (7) (2007), pp. 529–539
23. R. Sacks, M. Radosavljevic, R. Barak Requirements for building information modeling based lean production management systems for construction Automation in Construction, 19 (5) (2010), pp. 641–655
24. E. Gabriel The lean approach to project management International Journal of Project Management, 15 (4) (1997), pp. 205–209
25. A. Al-Sudairi, J. Diekmann, A. Songer, H. Brown, Simulation of Construction Processes: Traditional Practices versus Lean

Principles, University of California, Berkeley, CA, USA, Proceedings IGLC-7, 1999, pp. 39–50.

26. H. Thomas, J. Horman, R. Minchin, D. Chen Improving labor flow reliability for better productivity as lean construction principle Journal of Construction Engineering and Management, 129 (3) (2003), pp. 251–261
27. O. Salem, J. Solomon, A. Genaidy, M. Luegring Site implementation and assessment of lean construction techniques Lean Construction Journal, 2 (2005), pp. 1–21 ISSN: 1555-1369
28. A. Agbulos, Y. Mohamed, M. Al-Hussein, S. AbouRizk, J. Roesch Application of lean concepts and simulation analysis to improve efficiency of drainage operations maintenance crews Journal of Construction Engineering and Management, 132 (3) (2006), pp. 191–199
29. O. Salem, J. Solomon, A. Genaidy, I. Minkarah Lean construction: from theory to implementation Journal of Management in Engineering, 22 (4) (2006), pp. 168–175
30. X. Mao, X. Zhang Construction process reengineering by integrating lean principles and computer simulation techniques Journal of Construction Engineering and Management, 134 (5) (2008), pp. 371–381
31. L. Song, D. Liang Lean construction implementation and its implication on sustainability: a contractor's case study Canadian Journal of Civil Engineering, 38 (2011), pp. 350–359
32. S. Deshpande, E. Filson, O. Salem, R. Mille Lean techniques in the management of the design of an industrial project Journal of Management in Engineering, 28 (2) (2012), pp. 221–223
33. J. Shewchuk, C. Guo Panel stacking, panel sequencing, and stack locating in residential construction: lean approach Journal of Construction Engineering and Management, 138 (9) (2012), pp. 1006–1016
34. Construction Industry Institute (CII) Lean Implementation at the Project Level, vol. 234-1The University of Texas at Austin (2007) research summary, pp. 1–36
35. A. Alsehaimi, L. Koskela, Critical evaluation of previous delay studies in construction, in: Proceedings of the 8th International Postgraduate Conference, Prague, 2008.

36. A. Alsehaimi, L. Koskela, What can be learned from studies on delay in construction, in: Proceedings of the 16th IGLC Conference, Manchester, UK, 2008.

37. P. Love, I. Zahir, E. David Learning to reduce rework in projects: analysis of firm's organizational learning and quality practices Project Management Journal, 1 (September) (2003), pp. 13–25

38. Lean Construction Institute, What Is Lean Construction? <http://www.leanuk.leanconstruction.org/whatis.htm>, 2004.

39. L. Alarcon Tools for the identification and reduction of waste in construction projects Alarcon (Ed.), Lean Construction, A.A. Balkema, Rotterdam, The Netherlands (1994) 1997

40. C. Formoso, T. Soibelman, C. De Cesare, E. Isatto Material waste in building industry: main causes and prevention Journal of Construction Engineering and Management, 128 (4) (2002), pp. 316–325

41. C. Formoso, L. Soibelman, C. De Cesare, E. Isatto, Method of Waste Control in the Building Industry, Proceedings ICLG-7, University of California, Berkeley, CA, USA, 1999, pp. 325–334.

42. F. Picchi, A. Granja, Construction sites: using lean principles to seek broader implementation. 12th International Group for lean Construction Annual Meeting, Elsinore, vol. 1. 2004, pp. 1–1.

43. Construction Industry Institute (CII), Road Map for Lean Implementation at the Project Level, Bureau of Engineering Research, The University of Texas at Austin, Research Report 234-11, 2007.

44. O. Paez, J. Solomon, S. Salem, A. Genaidy Moving from lean manufacturing to lean construction: toward a common sociotechnological framework Wiley Periodicals, Human Factors and Ergonomics, Manufacturing Journal, 15 (2) (2005), pp. 233–245

45. G. Ballard Lean Project Delivery System Lean Construction Institute, Ketchum (2000)

46. A. Mossman Last Planner Overview: Collaborative Production Planning, Collaborative Programme Coordination Lean Construction Institute, UK (2005)

47. J. Auada, A. Scola, A. Conte, Last planner as a site operations tool, in: The 6th Annual Conference of the International Group for Lean Construction, (IGLC-6) Guaruj, Brazil, 1998.
48. L. Alarcon, Lean construction in Chile: a national strategy and local results, in: 3rd Annual Lean Construction Congress, <www.leanconstruction.org>, 2001.
49. J. Fernandez, V. Porwal, S. Lavy, A. Shafaat, Z. Rybkowski, K. Son, N. Lagoo Survey of motivations, benefits, and implementation challenges of last planner system users Journal of Construction Engineering and Management, 139 (4) (2013), pp. 354–360
50. G. Howell A Guide to the Last Planner for Construction Foremen and Supervisors Restricted White Paper Lean Construction Institute, California (2000)
51. B. Wambeke, M. Liu, S. Hsiang Using last planner and a risk assessment matrix to reduce variation in mechanical related construction tasks Journal of Construction Engineering and Management, 138 (4) (2012), pp. 491–498
52. Y. Kim, G. Ballard Management thinking in the earned value method system and the last planner system Journal of Management in Engineering, 26 (4) (2010), pp. 223–228
53. G. Howell, What is lean construction, in: Proceedings Seventh Annual Conference of the International Group for Lean Construction, IGLC-7, Berkeley, CA, 1999, pp. 1–10.

Chapter 6

Designing Bridges for Inspectability

Sreenivas Alampalli[1], and Arthur P. Yannotti[2]

[1]New York State Department of Transportation 50 Wolf Road, Albany, NY 12232 (518) 457-5498;

[2]New York State Department of Transportation 50 Wolf Road, Albany, NY 12232 (518) 457-6827;

ABSTRACT

Bridge inspection is essential for public safety and effective bridge management. Bridge inspection and bridge inspectors are the first line of defense against bridge failures. Most highway bridges in the United States and the rest of the world are predominantly inspected using visual methods by bridge inspectors manually assessing the condition of various elements. Thus, whenever possible, it is the responsibility of bridge designers to make sure that bridge details are designed such a way that the inspection of all the bridge components is visually possible. This paper highlights some of these issues and recommendations for designers based on the input from bridge inspectors working for the New York State Department of Transportation.

INTRODUCTION

Highway bridge inspections in the United States are mandated by the Federal Highway Administration (FHWA), as well as individual states, with procedures detailing inspection procedures, organization responsibilities, inspector qualifications, inspection frequency, reporting, and documentation. The original Highway Act passed by the United States Congress in 1916 that provided for federal aid to highways and subsequent highway acts include the requirements for maintenance of bridges as well as highways. Since about 1916, inspection of highway structures has been a part of maintenance work, although often a relatively minor part, by the states and other jurisdictions (White et al. 1992). More detailed inspections currently in place can be attributed to the Silver Bridge Collapse in 1967 and subsequent major failures. The present federal requirements can be found and are detailed in the FHWA National Bridge Inspection Standards (NBIS), last revised in 2004 and becoming effective in January 2005, sets minimum standards for the inspection of all publicly owned highway bridges (National 2004). These standards were set to ensure public safety by assuring that bridges have enough capacity to carry loads allowed on them. Inventory and certain condition data are also collected (Recording 1995) for assessing the condition of the entire bridge network in the nation and for determining allocations of federal highway bridge program funds (Alampalli and Jalinoos 2009). Many states and owners, including the New York State Department of Transportation (NYSDOT), collect more data for effective bridge management, i.e., decisions taken to support planning, design, maintenance, replacement, and rehabilitation activities. Some states also have more stringent requirements on personnel qualifications than those mandated by the NBIS. The current status of highway bridge inspections and use of NDE/NDT methodologies can be found in literature (e.g., Alampalli and Jalinoos 2009 and ASCE 2009).

ROLE OF BRIDGE DESIGN IN BRIDGE INSPECTION

Bridge life cycle is composed of several phases. These include planning, preliminary design, final design, construction, initial inspection following the construction, routine inspections throughout the service life of the bridge, maintenance, rehabilitation, and finally replacement. Many decisions taken during the planning and design may significantly influence the bridge service life as well as costs of inspection and maintenance. As noted earlier, the NBIS requires a full inspection of most highway bridges every twenty-four months and hands-on inspection of fracture critical components. New York State requirements are more stringent requiring hands-on inspection for all bridges and include component level inspection on a span basis. When certain conditions exist, New York State inspects bridges every twelve months or more often. New York State also specifies all Team Leaders to be Professional Engineers. Underwater inspections, generally performed by divers, are also required at least once every 60 months. Thus, these inspections do cost a considerable amount during the service life of the bridge. Inspectors also document the section loss and other deterioration so that critical findings can be documented and accurate load capacity of the structure can be evaluated to assure public safety (Alampalli and McCowan 2008). Thus, designers of new bridges or bridge rehabilitation must be cognizant of those requirements and take into consideration accessibility and costs of future bridge inspections. Inspection costs include, but are not limited to, personnel costs, access costs, work zone control costs, documentation costs, and evaluation costs. In some cases, the access and work zone control costs can be significantly higher than other costs due to heavy traffic volumes in urban areas. Thus, making access as easy as possible should be an important consideration in the bridge design process and has often been overlooked in the past. In addition, the use of details that make visual observation of critical components of primary members difficult should be avoided. The design codes have been relatively silent on these aspects even though AASHTO LRFR Bridge Design Specifications (AASHTO 2010) contain some requirements for bridge inspection that the designer needs to consider. The recent AASHTO Manual for Condition Evaluation of Bridges documents these requirements in more detail (The Manual 2008). Both of these manuals contain criteria for determining primary and secondary members and also contain criteria for further determining whether primary members are redundant or non-redundant and whether they are fracture critical or not. The NYSDOT Bridge Manual (Bridge 2010) also needs to be consulted for recommended design practices and details. The Bridge Manual contains suggested details for a variety of bridge types. Several aspects that can be considered during the design phase that can minimize inspection costs and improve inspection quality include: 1) bridge type selection, 2) type of details, and 3) access. These are discussed in the next sections. Some of the comments are based on responses received to a survey questionnaire that was sent in early 2010 to all inspection personnel within NYSDOT.

INSPECTION ISSUES: BRIDGE TYPE SELECTION

The most important element in designing for inspectability is the initial choice of a bridge type. In designing for inspectability, the first preference is to choose a structure that is load path redundant. This is an important consideration that affects inspection costs. The other types of redundancies are internally redundant and structurally redundant, but at present, the NYSDOT only recognizes load path redundancy in determining inspection requirements. A non-redundant structure that contains fracture critical members raises the level of necessary inspection even higher. Thus, a designer's first choice, from an inspection standpoint, should always be to choose a load path redundant structure. This is usually achieved by the selection of multi-girder bridges, irrespective of material used. By federal standards, any bridge with more than two girders is considered redundant. But, in New York State, to be considered load path redundant (Bridge Inspection Manual 1997), four or more girders are required. However, recent analytical studies indicate, that in specific cases, three and even two girder bridges can be load path redundant. Thus, in such cases, a designer documenting this on bridge plans and providing guidance to bridge inspectors can be very helpful. More documented guidance is needed in this area so that inspection costs can be reduced in the future. Other types of structures, such as multi-ribbed arches and concrete slab bridges, can also be considered load path redundant. However, sometimes load path redundant type structures cannot be used due to factors beyond the designer's control. These factors include site conditions and aesthetic concerns. One such example is the use of through trusses and through girders with limited vertical clearances or very long spans. In such circumstances, it is important for the designer to provide as much redundancy as possible, and furthermore, to include inspectable details and provisions in the design. Internal redundancy can be added to non-load path redundant structures. This is most important for fracture critical tension members. The tie girder in a tied arch can be made internally redundant by the use of multiple structural components connected by bolting rather than welding. Similar details can be used for truss chords or two girder bridges. In order for this internal redundancy to be effective, the designer must ensure that the remaining components of a fracture critical member have sufficient capacity to prevent collapse of the structure after the failure of an individual component. Steel box or tub girders can be attractive structural types and are especially efficient for curved structures because of their high torsional strength. They are, however, inherently more difficult to inspect than I-shaped girders. The designer can improve the inspectability by making them large enough for access. A minimum depth of 5 feet 6 inches is recommended. In addition, the diaphragm openings need to be large enough for an inspector to pass through. Inspectability can also be enhanced by painting the interior white and providing interior lighting (see Figure 1).

Figure 1: Inspection access for inside the box sections.

Adjacent prestressed concrete box girders have been a common and favored structural choice. They are an economical structure and have the advantage of a small structure depth where vertical clearance is an issue and having a smooth underside that helps to prevent the snagging of debris carried by waterways. They do, however, present inspection challenges. Prestressed girders, by their nature, are harder to assess visually. In addition, adjacent box girder design prevents the bridge inspector from making a visual assessment of the sides of the girder.

INSPECTION ISSUES: BRIDGE DETAILS FOR INSPECTION

Access Bridge details play a major role in ensuring bridge safety as accessibility to all components of a bridge is important. If a component cannot be reached or is hidden to the inspector, the inspector cannot examine the component and cannot determine its condition to assure that it is behaving as designed. Sometimes it is not possible to make all components directly visible to the inspector, but provisions should be made that simple devices available to the inspector (eg. mirror) can be used to assess their condition.

Abutments and Piers

One of the common issues faced by bridge inspectors is accessibility to the abutments and piers. In most cases, these can be dealt with relatively easily during the bridge design. Some details that can assist inspectors include the following:

- When the exposed face of the abutment stem is four feet or higher, providing a flat spot or berm in front of the abutment stem will allow a ladder to be placed for easy access to the bridge seat and bearings.
- During bridge replacement projects, new abutments are often built behind the old abutments with old abutments left in place for slope/scour protection. In such cases, removing enough of the old abutment to provide at least a three feet clearance to the new superstructure above will give bridge inspectors access to the new abutments.
- When it is anticipated that access to the piers will be by boat, provisions for tying off the inspection boat to the pier will be of use to bridge inspectors. If it is anticipated that use of the Under Bridge Inspection Unit (UBIU) is not possible to inspect piers in water due to the deep superstructure or other design details, a provision for inspection access ladders on the piers should be considered.

Trusses and Arches

Truss chords and arches commonly use box sections to minimize debris accumulation. However, their closed section makes visual inspection difficult and thus, providing periodic access ports of sufficient size on the underside of the box should be considered to provide inspection access. Removable screens should be provided with these access ports to prevent the ingress of nesting birds. Large through and deck trusses are normally used over waterways to accommodate required long spans. In such cases, UBIUs are commonly used for inspection. Hence, openings between the truss verticals and diagonals should be large enough to accommodate the passage of the UBIU bucket. Designers should consult with bridge inspection personnel to determine the size of the equipment typically used. The openings between the verticals and diagonals are most easily controlled by setting the panel length as long as possible to maximize the opening. Similarly, an effort should be made to accommodate sidewalks inside the trusses to make inspection access by UBIU relatively easier (see Figure 2). Utilities should be provided inside the truss whenever possible to make access with the UBIU easier for inspection of the bridge's underside (see Figure 3).

Figure 2: Bridge with sidewalks outside the truss.

Figure 3: Bridge with utilities outside the truss.

Girder Bridges

I-girder bridges are commonly used and are also the easiest type of bridge to inspect since most elements are visually accessible to the inspector. But, whenever possible, Class D, E, and E′ fatigue details should be avoided (AASHTO 2010) as these necessitate special emphasis inspection. In such cases, use of bolted connections of all cross frame and laterals is recommended while, in general, avoiding complex details. Stay-in-place forms are more economical than removable forms from a construction perspective and thus are commonly used in New York State. However, their use makes the visual inspection of the bridge deck more difficult as illustrated in the Figure 4. In such cases, removal of some panels should be considered for deck inspection.

Figure 4: Stay-in-place forms.

Inspection hand rails are commonly provided on girder bridges even though their use is debated by several inspectors (see Figure 5). When hand rails are provided, they must be robust enough to safely support a bridge inspector. Current NYSDOT practice is to provide inspection hand rails on girders having a web depth of 5 feet or greater on both sides of interior girders and on the inside of fascia girders (Bridge Manual 2010).

Figure 5: Use of handrail in deep sections.

End Diaphragms

Almost all girder bridges have a line of end diaphragms to support the deck slab at joints and to convey lateral loads to the bearings and substructure. If solid diaphragms are used at the abutments, a sufficient opening should be provided to permit the bridge inspector access to the bearings and abutment back wall.

Bridge Railing and Fencing

Fencing is frequently provided on bridges over roadways and railroad tracks to discourage vandalism to traffic beneath the bridge. However, fencing can make inspection access by UBIU difficult. Although not commonly used, it should be possible to provide sections of fence with swing down details to accommodate UBIU access. Of course, any such detail should be provided with secure locking mechanisms.

Bridge Inspection and Security

Bridge security is becoming a very important aspect and consideration of a bridge's security is now part of the bridge design process in most states. Thus, protecting critical components of bridges is routinely undertaken by bridge owners, especially in urban areas. One popular method used is to make sure the critical components are concealed by protective coverings, but this can make inspection operations and access very difficult. This can also increase the inspection costs if the protective coverings have to be dismantled and reassembled during inspection. Careful consideration for inspection of concealed components should be given before such protective measures are undertaken.

Recent and Other Ongoing Activities at NYSDOT

The current NBIS (2004) requires that specialized inspection procedures be identified for "complex bridges." It defines potentially complex bridges as "moveable, suspension, cable stayed, and other bridges with unusual characteristics." In order to address this requirement as well as to provide guidance to bridge inspectors when uncommon or unique details are used by the designers, a recent update of the Bridge Manual (2010) included the following language to alert designers. Note that every bridge in New York State has a unique bridge identification number (BIN) and all inspection related documentation for each bridge is placed in the appropriate BIN folder. The designer of innovative or complex bridges, or of bridges that incorporate innovative or unusual elements or details, shall identify those bridge elements or details that warrant specialized inspection attention. A "Special Emphasis Inspection Procedure (SEIP)" shall be assembled, submitted with the "Final Bridge Plans," and stored in the BIN file. The contents of the SEIP shall be as follows: A brief statement of purpose presenting an explanation of why the SEIP is required. Characterization of the complex of unusual elements or details to be given special attention; a description of the element or detail and the reason(s) it is considered complex or unusual. A description of how the element or detail should function or behave along with a description of the physical conditions that can be observed that would indicate that the element or detail is functioning appropriately. A description of the observable physical characteristics that would indicate that the element or detail is not functioning appropriately along with direction relative to how to assess the degree to which the element or detail is faulty in its performance. When new materials (such as fiber reinforced polymers) or new design details are used, the designer should provide guidance to bridge inspection personnel by incorporating appropriate comments on the final plans. This could also include suggestions for periodic load testing or the use of instrumentation for periodic structural health monitoring based on cost-benefit analysis. In such cases, the designer should identify on the plans the properties that should be measured, locations for measuring, what items should be given attention during routine inspections, etc. The bridge selection and detailing is mostly done by experience. Cost-effectiveness is given a priority during the design and is taught at colleges in design courses. In most cases, inspectability is not discussed unless designers have some inspection experience. No written guidelines exist for designers. Hence, NYSDOT is considering preparing guidelines on designing for inspectability and providing this to all the designers.

SUMMARY

A properly designed bridge is one that accounts for bridge inspection and maintenance. Bridge inspection is not routinely considered during the bridge design process and this paper discusses designing bridges for inspectability. Selection of bridge type, component details, access requirements, and effect of considering security in recent years, and other aspects are discussed based on the input received from bridge inspectors working for the New York State Department of Transportation.

ACKNOWLEDGEMENTS

The opinions expressed in this paper are those of the authors and not necessarily of the organization they represent. Authors also acknowledge the regional bridge management groups and Abate Associates Engineers and Surveyors for providing input on this topic.

REFERENCES

1. AASHTO LRFD Bridge Design Specifications, American Association of State Highway and Transportation Officials, Washington, D.C., 2010
2. Alampalli, S. and McCowan, P. "Addressing Critical Findings from Highway Bridge Inspection: New York State Procedures," Proceedings of Structural Materials Technology (SMT): NDE/NDT for Highways and Bridges Topical Conference, American Society for Nondestructive Testing, Oakland, CA, September 2008
3. Alampalli, S. and Jalinoos, F. "Use of NDT Technologies in US Bridge Inspection Practice," Materials Evaluation, Journal in Nondestructive Testing/Evaluation/Inspection, American Society for Nondestructive Testing, 67(11), 1236-1246, 2009
4. ASCE, ASCE/SEI-AASHTO Ad Hoc Group on Bridge Inspection, Rating, Rehabilitation, and Replacement, "White Paper on Bridge Inspection and Rating," Journal of Bridge Engineering, 14(1), 1–5, 2009
5. Bridge Inspection Manual, New York State Department of Transportation, Albany, NY, 1997
6. Bridge Manual, New York State Department of Transportation, Albany, NY, April 2010
7. Recording and Coding Guide for the Structural Inventory and Appraisal of the Nation's Bridges, FHWA-PD- 96-001, Washington, Federal Highway Administration, December 1995
8. National Bridge Inspection Standards, 23 CFR Part 650, Federal Highway Administration, US Department of Transportation, Federal Register, Vol. 69, No. 239, 14, December 2004
9. The Manual for Bridge Evaluation, American Association of State Highway and Transportation Officials, Washington, D.C., 2008 (with 2010 interim Revisions)
10. White, K. R., Minor, J., and Derucher, K. N. "Bridge Maintenance Inspection and Evaluation," Marcel Dekker, Inc., New York, 1992

Chapter 7

Probabilistic Evaluation of Service Life for Reinforced Concrete Structures

Sanjeev Kumar Verma[1], Sudhir Singh Bhadauria[2], and Saleem Akhtar[1]

[1]Civil Engineering Department, University Institute of Technology, Rajiv Gandhi Technological University, Airport Road Bhopal, Madhya Pradesh 462036, India.

[2]S.G.S. Institute of Technology and Science, Indore, Madhya Pradesh 452003, India.

ABSTRACT

Degradation of performance and deterioration of different components of reinforced concrete (RC) structures increase with the age of structure. This deterioration of reinforced component depends on several parameters. However, modeling service life of RC structure by considering all the parameters is a difficult job, as most of the parameters are uncertain in nature. Probabilistic models account well for the uncertainties in the parameters responsible for deterioration of RC structures. This paper presents a review of several recent service life models developed using probability based concepts.

INTRODUCTION

Performance and service life of a RC structure are governed by several parameters such as strength, quality of concrete, concrete cover, age, and most significantly by exposure conditions. According to Li et al. [1] behavior or performance of concrete structures depends on various physical and chemical related phenomena but it is hard to combine various analytical models in order to simulate real degradation process. Hence, because of random nature of the parameters governing the performance of RC structures, probabilistic approach is better to develop reliable service life prediction model. Probabilistic models calibrated with monitored field data can provide more reliable models to predict the probabilities of corrosion. This category includes models based on probability or statistical concepts such as probability function, reliability based models, Monte Carlo simulation, Markov chain methods, and fuzzy logic. In probablistic analysis three steps can be distinguished: (a) definition of failure modes (limit states) and corresponding models, (b) quantification of the statistics of the random variables, and (c) calculation of the desired results as, for instance, failure probabilities [2]. Present paper reviewed several probability based service life models developed by researchers for evaluating the probability of failure and estimating the residual life of RC structures.

SERVICE LIFE

Service life of a RC structure can be defined as time period during which structure will fulfill the required performance under defined repair and maintenance. Zhang and Ba [3] defined service life as the life before critical chloride concentration led to reinforcement corrosion. According to Chai et al. [4] period from initial use to depassivation can be considered as service life. Vorechovska et al. [5] considered depassivation of reinforcement due to carbonation or chlorination as the end of service life. Parameswaran et al. [6] considered the time during which the performance of a concrete structure will be kept at a compatible level to fulfill the performance as service life, provided that it is properly maintained. Rostam [7] defined service life as the number of years during which the structure shall perform satisfactorily without unforeseen high costs for maintenance. Martin-Perez and Lounis [8] defined service life as the time until damage accumulation reaches an unacceptable or limit state. According to Abu-Tair et al. [9] end of service life is reached when either failure occurs or major repair is necessary to keep the structure in service. Teply [10] described service life as the period of time after installation during which all the conditions of the structure meet or exceed the performance. Different levels in service life of a RC structure are presented in Figure 1; these levels are design, construction, in use, period without any deterioration, initiation of deterioration, propagation of deterioration, formation of cracks, rapid deterioration, and finally failure of structure.

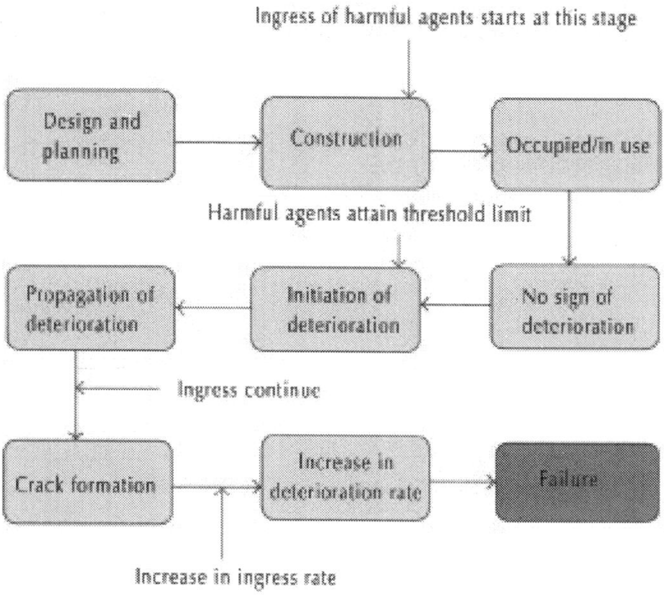

Figure 1: Different levels of service life.

According to Figure 1, after design and construction when structure is in use before the initiation of deterioration, there is a phase with no sign of deterioration. Ingress of harmful agents will get initiated during the construction level and continuous during no deterioration level. When ingress of theses harmful agents reaches the threshold limit then it initiates deterioration of RC structures; this deterioration further propagates and reduces the performance of RC structure. Deterioration propagates and finally results in formation of cracks; these cracks allow more ingress of harmful agents and further increases the deterioration at a faster rate. Finally this increase in deterioration rate reduces the performance and finally causes the failure of structure.

For modeling the performance of RC structures several parameters such as environmental effects, degradation mechanisms, durability parameters, and properties of ingredients are required to be considered. Defining the failure of structure is most important for modeling the service life. Failure of a concrete structure can be simply explained on the basis of load "L" and capacity "C" L and C are any destructive effect and resistive capacity. If L is the environmental effect, C is the capacity to resist the environmental effects. Hence if L is greater than C, this condition will represent the failure of structure "F":

$$\{F\} = \{C < L\} \quad (1)$$

Therefore probability of failure Pf can be defined as

$$\{F\} = \{C < L\} \quad (2)$$

SEVERAL PROBABILISTIC PERFORMANCE MODELS

Probabilistic models account well for the uncertainties in the parameters responsible for deterioration of RC structures. Hence, several researchers applied probability concepts for developing performance models. Orcesi and Frangopol [11] implemented a model using the lifetime functions to evaluate a probability of survival of bridge components. The possible outcomes assossiated with nondestructive inspections (NDIs) are incorporated in an event tree model. Use of life time function for each component of the structural system enables to express the probability that the component survives and also provided a practical methodology for determining optimal NDIs for different components of steel bridges. The expression of the survival function (S) is shown in the following equation:

$$S(t) = 1 - F(t) = P(T > t) = \int_{t}^{\infty} f(u)du \qquad (3)$$

where S is the lifetime function that represents the probability that structure will not fail before time t. F is the represents the probability that the structure fails before time t. T is the time to failure. f is the probability density function. Cusson et al. [12] presented the development of probablistic mechanistic modeling approach supported by durability monitoring to obtain improved predictions of service life of concrete bridge decks exposed to chlorides. Factors influencing the service life are (i) surface chloride content C_s, (ii) chloride diffusion coefficient D, (iii) threshold chloride content C_{th}, (iv) corrosion rate and deterioration rate.

Agarwal et al. [13] described an approach based on Weibull distribution to calculate the deterioration rates of typical bridge elements using historical bridge inspection data and compare the results with those, using the traditionally used Markov chains approach. This approach has been incorporated into a computer program that generates the deterioration curves for bridge elements, based on historical bridge inspection data and described a 7-scale condition rating used to calculate the deterioration rates for each bridge element. Bridge condition rating scale ranges from 7 to 1, 7 being new and 1 being failed condition. Mitra et al. [14] developed method for obtaining condition index of corrosion distressed RC building using concepts of fuzzy logic and integrates visual inspection with in situ investigations for carbonation and chloride content. Visual assessment is a two stage process: (i) visual survey has been conducted to gain information about physical and chemical state and by recording visible signs of deterioration such as cracks, rust stains, and spalling colour; (ii) condition has been assessed by an expect based on the information obtained in first stage. Vorechovska et al. [5] presented the service life assessment utilizing durability limit states specialized for concrete structures. Both initiation and propagation periods of reinforcement corrosion are considered and a comprehensive choice of limit states is provided. The approach is based on degradation modelling and probablistic assessment.

Strauss et al. [15] introduced a reliability assessment of RC structures and demonstrated real existing bridge structures. Reliability index of the structure decreases during its life cycle due to material degradation. This procedure is modeled by advanced life cycle computer simulation. Analytical deterioration models combined with in situ monitoring are used to launch the degradation process. Main feature of the presented approach is the nonlinear finite element analysis of the structure. Marano et al. [16] developed a probability based methodology to use the traditional theories for reliability assessment and proposed an efficient approach for time-dependent reliability analysis regarding RC beams subjected to pitting corrosion. Steel bars pitting corrosion has been an electrochemical process that could seriously degrade the RC structures. Consequences of pitting corrosion process on structural safety are more serious than due to general corrosion. Tesfamariam and Martín-Pérez [17] proposed a probablistic based model for carbonation induced corrosion based on Bayesian belief network (BBN); this proposed model is compared with more traditional probablistic technique, Monte Carlo simulation. This BBN model estimates carbonation depth and consequently determines the probability of onset of corrosion. Commonly available information of environmental exposure condition, mix design information, and in situ nondestructive testing measurements have been integrated to determine the potential for carbonation indused corrosion. For carbonation induced corrosion, service life is the time for the carbonation front to reach the reinforcing steel depth. Relation between carbonation depth and exposure time is given in the following equation:

$$x = kt^{(1/m)} \tag{4}$$

where x is carbonation depth, k is carbonation constant, m is time of exposure, normally is 2. Sobhani and Ramezaninpour [18] developed an algorithm to extract the fuzzy membership functions from the available stochastic information and also proposed an integrated system to convert the probablistic information into the corresponding fuzzy sets. Under environmental conditions of high aggressiveness, even concrete which has been properly prepared and placed may loose its protective properties and allow for corrosion of reinforcement. Corrosion of reinforcement involves some uncertainties: (1) random nature of corrosion process and (2) linguistic terms of construction process.

Melchers et. al [19] proposed an approach for strength and performance deterioration modeling of RC structures exposed to saline or marine environment, based on theoretical concepts and probability. At present there is little field experience and other data available, so probablistic modeling has been proposed here. This has been based on the evidence that the rate of diffusion of chlorides is influenced by internal damage to the concrete surrounding the reinforcement. The proposed procedure has been applied to an example beam and compared to experimental observations, including estimates of uncertainty in the remaining ultimate moment capacity and beam stiffness. Val [20] examined the effect of corrosion of reinforcing steel on flexural and shear strength and on the reliability of reinforced concrete beams. Probability of failure has been evaluated using Monte Carlo simulation. Uncertainties in material properties, geometry, loads, and corrosion modeling are taken into account. Two types of corrosion general and pitting have been considered and more emphasis has been on pitting corrosion. Results show that pitting corrosion has a significant influence on the reliability of RC beams. Anoop and Balaji Rao [21] demonstrated that the use of data from field inspections for performance evaluation and remaining life assessment of corrosion affected RC bridge girders using the concept of additive fuzzy logic; remaining life of the member can be assessed by determining the time taken for a given performance measure to deteriorate to a target value. Corrosion initiation time is determined using Fick's 2nd law of diffusion. Uncertainties in the values of the parameters characterizing the environment and variables affecting the time to corrosion initiation and corrosion propagation are taken into account by treating them as fuzzy variables. Usefulness of proposed methodology has been illustrated through a case study, by comparing the time to reach different damage levels for a severely distressed beam. Shao and Li [22] formulated an asset management strategy based on risk cost optimization for infrastructure during its whole service life. Time-dependent reliability method is used to determine the probability of attaining each phase of the service life. An algorithm is developed for above formulation and programmed in a user-friendly manner. To maintain the safety and serviceability of a structure during its service life, intermediate maintenance for the structure has been necessary. Li and Vrouwenvelder [2] showed the service life prediction procedure including advanced spatial variability approach, which implements the random spatial variation of property differences across the structure. Most parameters involved in deterioration of RC structures subjected to corrosion are random, so probability based reliability analysis is often applied. In probablistic analysis three steps can be distinguished: (a) definition of failure modes (limit states) and corresponding models, (b) quantification of the statistics of the random variables, and (c) calculation of the desired results as, for instance, failure probabilities. Chloride ingress is modeled using Fick's 2nd law of diffusion.

Sasmal et al. [23] explored the possibilities of using fuzzy mathematics for condition assessment and rating of bridges and developing a systematic procedure and formulations for rating existing bridges using fuzzy mathematics. Fuzzy logic is a means for modeling the uncertainty involved in describing an event result using natural language; it is useful for remedying the uncertainties and imprecision. Computer programs developed from formulations presented in this paper have been used for evaluating the rate of existing bridges.

Li et al. [24] presented a performance based methodology for serviceability assessment of corrosion affected concrete structures. Methods of time-dependent reliability are employed to quantify the probability of serviceability failure. Merit of the developed methodology has been identified as the structural response has been directly related to design criteria used by structural engineers and asset managers. It has been found that, under the same service conditions, corrosion-induced cracking attains a serviceability limit state before deflection does. In the theory of structural reliability the criterion can be expressed in the form of a limit state function as follows:

$$G(L,S,t) = L(t) \tag{5}$$

where S(t) denotes the response of structure at time t, such as stresses and deformation; and denotes a critical limit for structural response, which may change with time but in most practical use it is a constant prescribed in the design codes. With the limit state function of (5), the probability of serviceability failure, ps, can be determined by

$$Ps(t) = P\left[G(L,S,t) \le 0\right] = P\left[S(t)\right] \quad (6)$$

Kong and Frangopol [25] simulated time-varying performance of deteriorating materials and structures using probabilistic methods. Evaluation of the time-varying performance of deteriorating materials and its effect on structures measured by the reliability index profile is addressed based on three aspects: (1) acquisition of statistical characteristics of input random variables from imperfect samples, (2) model-based simulation, and (3) evaluation of variability and sensitivity measures of the nondeterministic performance model. Vu and Stewart [26] developed a 2D spatial time-dependent reliability model to predict the likelihood and extent of corrosion-induced cracking. The spatial variability of concrete cover, concrete compressive strength, and surface chloride concentration is considered in the spatial time-dependent reliability model. The reliability analysis predicts (1) probability of the first incidence of cracking, (2) proportion of an area subject to severe cracking, and (3) probability that a given percentage of a concrete surface has cracked. The time to severe cracking has been referred to the time when concrete cover cracking reaches a limit crack width and has been estimated based on results from accelerated corrosion tests of RC slabs. The corrosion-induced cracks occurred on the concrete surface above and parallel to the rebars. The various stages of crack growth can be represented in Figure 2, where time to excessive cracking can be described in two stages: (1) crack initiation , time to first cracking (hairline crack of 0.05 mm width), (2) crack propagation (t_2); time for crack to develop from first cracking to a limit crack width.

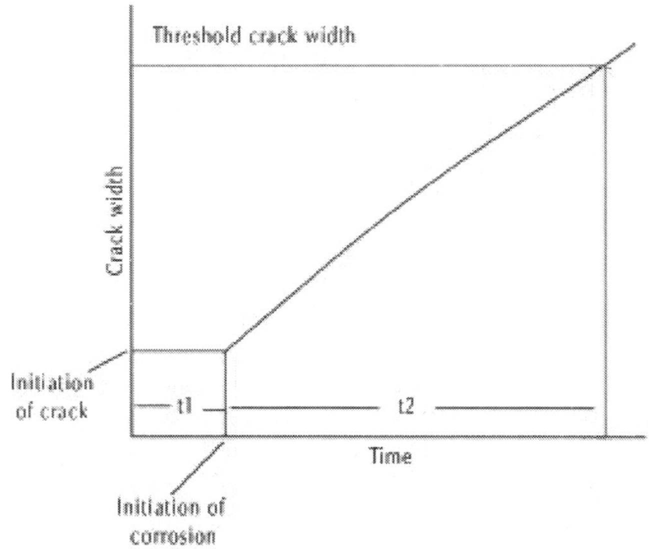

Figure 2: Crack initiation and propagation based on concept of Vu and Stewart [26].

Akgül and Frangopol [27] presented a probablistic framework forecasting the lifetime performance of RC slabs and girders of bridges. Bridges have been individually described followed by presentation of the performance analysis results including both initial reliability indices and lifetime reliability profiles. Computer program RELNET (reliability of system network) has been used to perform reliability analysis for the bridges. Special emphasis has been given to chloride penetration in RC. Use of Fick's 2nd law of diffusion for modeling the chloride penetration has been explained. Val [28] presented a reliability-based method for the evaluation of the expected cost of failure of RC structures subject to chloride-induced corrosion, and penetration of chloride into concrete is studied by Fick's 2nd law of diffusion. The resistance degradation function introduced has been used to calculate the probability of failure of corrosion-damaged RC elements for such typical ultimate limit states as flexure, shear, and compression. In this study only the most essential parameters, which govern the deterioration process associated with corrosion, have been considered. These include concrete grade and thickness of the concrete cover (structural parameters) and the surface chloride content (environmental conditions).

Martin-Perez and Lounis [8] presented an approach for service life prediction of RC structures exposed to chloride environment that combines a FE modeling of the chloride transport and a reliability-based analytical method for onset of damage and its accumulation. By using Monte Carlo simulation the probabilistic distributions of the chloride penetration front and corrosion initiation time are generated. A stochastic deterioration process for corrosion initiation and propagation and then crack initiation and propagation have been used to examine the effect of cracking, spalling, and loss of reinforcement area on structural strength and reliability by Stewart and Val [29] and also presented models for reliability and life cycle cost analyses of reinforced concrete bridges damaged by corrosion. Monte Carlo simulation analysis is then used to determine probabilistic time-dependent estimates of structural performance, such as resistance and reliability of serviceability and strength limit states, for a typical RC slab bridge. In the present analyses it is assumed that corrosion, dimensional, and material variables are homogeneous over the structure so spalling damage will also be initiated homogeneously over the deck area. Li [30] proposed a performance-based model of life cycles of corrosion affected RC structures with the focus on the effect of propagation of reinforcement corrosion on structural deterioration, corrosion propagation that leads to structural deterioration which is of most concern to structural engineers. Results produced from a comprehensive experiment designed to investigate the corrosion propagation in concrete and its effect on the deterioration of structural strength and stiffness have been presented and analyzed. An assessment criterion has been established to define the structural performance in terms of strength and serviceability limit states that are consistent with design criteria. A comparison of the experimental results on strength deterioration as determined from destructive load test and nondestructive measurement of corrosion current density has also been undertaken. In addition, the proposed model is on a probabilistic basis. A performance-based assessment criterion needs to be established:

$$P\left[R(t) \geq Ra\right] \geq pa \qquad (7)$$

where P is probability of an event and R(t) is structural resistance varying with time t, that is, deterioration; Ra is minimum acceptable resistance and pa is minimum acceptable probability of structural safety, that is, structural reliability. Detection of the initiation of corrosion in steel bars denotes the beginning of whole life performance assessment for RC structures. Li [31] emphasize on the development of a reliability-based method to estimate the initiation time of reinforcement corrosion in concrete structures. Detection of the initiation of corrosion in reinforcement marks the beginning of whole life performance assessment for RC structures. Failure does not necessarily mean structural collapse only, but also includes loss of serviceability, characterized by concrete cracking, spalling, and excessive deflection. Merit of developed models is that they are derived from data representing concrete structures in real service conditions.

Liang et al. [32] adopted multiple layer fuzzy theory to evaluate the state of existing RC bridges. The damage state of any member of an existing bridge has been systematically and completely composed as an evaluation model of multiple layer fuzzy mathematics. The evaluated results may be used for the safety index and reference index for repair or reinforcement in existing bridges. In addition, the evaluated results may also be used as a design reference for service life in future bridge building. The evaluated bridges have been categorized into the grades I, II, III, IV, and V, which are described as nondamage, light damage, moderate damage, severe damage, and unfit for service, respectively. Teply [10] applied a probabilistic approach to predict the service life of concrete structures for modeling the deterioration effects. Mathematical models can be made for the performances of a building or structure, including the process of ageing (deteriorating). Generally two classes of problems may be distinguished in this context: (i) prediction of service life for structures during the design process. This becomes a necessity in the context of the new "performance-based design" rules, which are a trend in forthcoming engineering work; (ii) assessment of existing structures, that is, estimating residual service life. Figure 3 shows the probabilistic representation of service life comparing influence of the load S and a function for the load bearing capacity R (or the structural resistance function). Alternatively, the service life of reinforced concrete structures may be assessed by distinguishing two periods. First, the initiation period, during which the gradual depassivation of the reinforcing bars (due to CO_2 or other effects) is in progress. Second, the propagation period, during which corrosion of the reinforcement takes place, and, consequently, the failure probability is increased for both the load and serviceability limit states; this is illustrated in Figure 4.

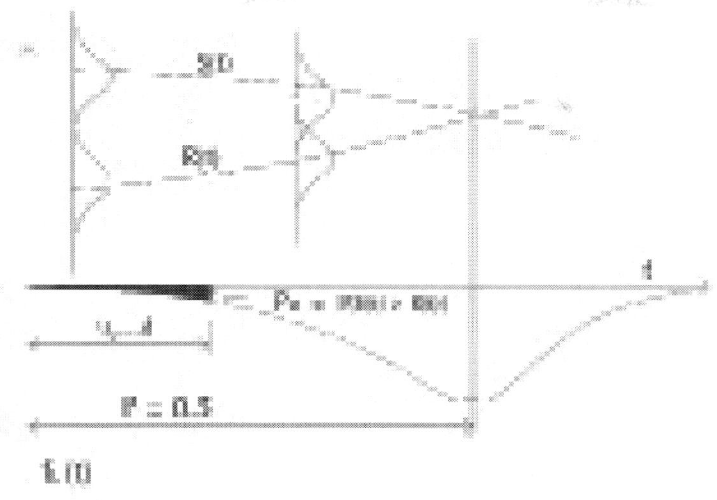

Figure 3: Probabilistic structural service life by Teply [10].

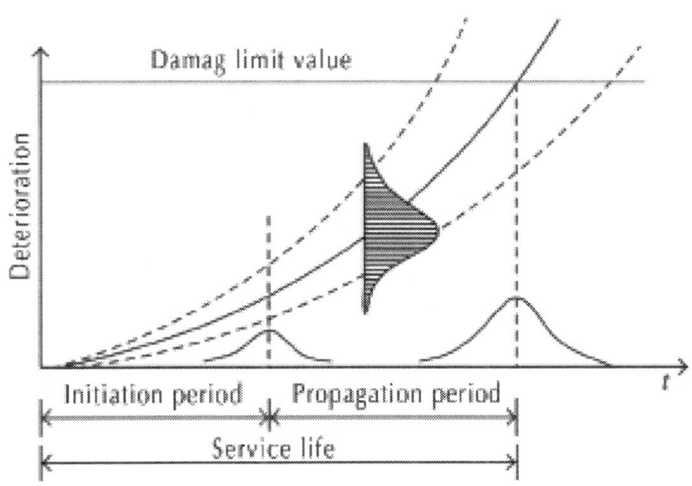

Figure 4: Two periods of structural service life by Teply [10].

Gulikers [33] presented a model which takes into account the accelerating effect of macrocells on local corrosion rates assuming a cylindrical shape of the pitting site; results are presented in terms of local penetration depth, surface area under corrosion attack, and steel volume consumed by corrosion. The improved corrosion deterioration model can be used to develop a probabilistic model which allows for evaluation of the probability that a given reduction of the reinforcement cross-section has taken place at a given time. Abu-Tair et al. [34] highlighted the importance of developing service life prediction models for the risk of degradation of reinforced concrete structures so that optimal strategies for their maintenance and repair can be developed; data on bridge deterioration of over 400 bridges will be used to provide predictive models based on the factor method. Vu and Stewart [35] measured service life of structures exposed to aggressive environments by the probability of cracking and spalling of concrete cover; time to corrosion/spalling is found experimentally from accelerated corrosion testing of RC slabs, emphasis to quantify the relationship between concrete quality (w/c ratio or strength), concrete cover, crack propagation, and time. Proportion of a concrete surface likely to spall for any reference period can be predicted by reliability model. This is a useful criterion for predicting the service life of RC structures.

The MCSC (Monte Carlo simulation for corrosion) computer program has been used to evaluate the probability density function of the corrosion initiation time for the steel reinforcement by Enright and Frangopol [36], for the bridge girders with particular attention to Colorado bridges. The corrosion initiation time is distributed log normally and is increasing with the depth of concrete cover. A survey on degradation mechanisms is briefly presented. Damage has been reviewed for a variety of bridges based on information collected from the literature and from field studies performed by the Department of Transportation, Colorado. The scope of this investigation has been limited to concrete bridges subjected to environmental attack. Method for predicting strength loss has been applied to a typical bridge in Colorado. The results can be used to identify critical elements for inspection and repair and to assist in the development of rational maintenance planning strategies for concrete bridges. The strength degradation mechanisms can be classified as affecting either the concrete or the steel reinforcement or both. Concrete deteriorates because of internal pressures, which have been caused primarily by chemical reactions in the cement (sulfate attack), by chemical reactions between the cement and aggregate (alkali-silica reaction), or by freeze-thaw cycle attack.

According to Naticchia [37] building behavior depends on various physical and chemical phenomena related to each other, but it is hard to combine many numerical models in order to simulate real behavior of degradation process. In order to find ideal strategy, a model is proposed based on "qualitative simulation approach" capable to represent the interrelated decay phenomena, so an approach is proposed based on Bayesian belief networks, assembling all phenomena. Scheremans and Gemert [38] updated a probability based model for interpretation of test results and prediction of the service life of RC structures. Ingress of chloride ions into the pores of the concrete is caused by diffusion and evaluated using Fick's 2nd law of diffusion. Teply et al. [39] described a probabilistic approach for investigating the reliability and predicting life time of simply supported RC beams including deterioration in time. Only the ultimate limit state is controlled and depassivation of reinforcing bars is supposed to be caused by carbonation of concrete. Enright and Frangopol [40] combined inspection results with prior information using Bayesian updating and used this approach for determining corrosion initiation time and rate on time-variant strength. This study illustrated a method for computing strength degradation using a probabilistic approach, updated by Bayesian's method.

Cheung and Kyle [41] described the development and application of reliability-based system for the service life prediction and management of maintenance and repair procedures of RC structures. This system is capable of establishing levels of reliability by using statistical databases and probability theory, for satisfying various governing conditions (limit states) required for determining performance and financial considerations. The reliability of a structure or component has been defined as its probability of survival (Ps), which is related to probability of failure (Pf) as in (8): $Ps = 1 - Pf. (8)$

Mori and Ellingwood [42] presented a probability based method to evaluate time-dependent reliability of components and systems of reinforced or prestressed concrete structures. These methods enable the impact of uncertainties in loading condition, structural strength, and strength degradation on safety and serviceability of structures due to aggressive environment. The role of periodic inspection and maintenance in enabling a target reliability level to be met over a period of continued service has also been considered.

DISCUSSION AND CONCLUSIONS

In most of the above models, service life of RC structures is divided in different phases. First conceptual model of service life has been provided by Tuutti in 1982. Tuutti's model divides service life in two phases: first phase represents the time required by chloride ion or carbonation to initiate the corrosion and second phase represents the time after corrosion initiation until structure fails.

From the above review it has been observed that Bayesian belief network models are simple to use and required input data can be easily obtained from the user. Fuzzy logic approach is the best for removing the uncertainties and imprecision in observed values. Markov chain models are useful when very few or no inspection data is available on severely deteriorated structures; however, it cannot account for unobserved durations and regular inspections of real structures are required for performing time-dependent reliability analysis. Semi-Markov process is one of the best ways to interpret the deterioration of structures subjected to harsh environments, as they consider different damage levels of the structures with different service lives. Weibull-based approach performs better in evaluating probability of failure.

Probabilistic models account well for the uncertainties in the parameters responsible for deterioration of RC structures. Therefore, because of the random nature of the parameters governing the performance of RC structures, probabilistic approach is better to develop reliable service life prediction model. Probabilistic models calibrated with monitored field data can provide more reliable models to predict the probabilities of corrosion.

REFERENCES

1. Z. Li, C. K. Chau, and X. Zhou, "Accelerated assessment and fuzzy evaluation of concrete durability,"Journal of Materials in Civil Engineering, vol. 17, no. 3, pp. 257–263, 2005

2. Y. Li and T. Vrouwenvelder, "Service life prediction and repair of concrete structures with spatial variability," Heron, vol. 52, no. 4, pp. 251–267, 2007
3. W.-M. Zhang and H.-J. Ba, "Accelerated life test of concrete in chloride environment," Journal of Materials in Civil Engineering, vol. 23, no. 3, pp. 330–334, 2011.
4. W. Chai, W. Li, and H. Ba, "Experimental study on predicting service life of concrete in the marine environment," Open Civil Engineering Journal, vol. 5, no. 1, pp. 93–99, 2011
5. D. Vorechovska, B. Teply, and M. Chroma, "Probabilistic assessment of concrete structure durability under reinforcement corrosion," Journal of Performance of Constructed Facilities, vol. 24, no. 6, pp. 571–579, 2010.
6. L. Parameswaran, R. Kumar, and G. K. Sahu, "Effect of carbonation on concrete bridge service life,"Journal of Bridge Engineering, vol. 13, no. 1, pp. 75–82, 2008.
7. S. Rostam, "Service life design of concrete structures: a challenge to designers as well as owners,"Asian Journal of Civil Engineering, vol. 6, no. 5, pp. 423–445, 2005.
8. B. Martin-Perez and Z. Lounis, "Numerical modeling of service life of reinforced concrete structures," in Proceedings of the 2nd International RILEM Workshop on Life Prediction and Aging Management of Concrete Structures, pp. 71–79, Paris, France, May 2003.
9. A. Abu-Tair, C. McParland, J. F. Lyness, and A. Nadjai, "Predictive models of deterioration rates of concrete bridges using the factor method based on historic inspection data," in Proceedings of thof the 9th Durability of Building Materials and Components (DBMC '02), pp. 2621–2629, Brisban, Australia, 2002.
10. B. Teply, "Modelling of deterioration effects on concrete structures," Acta Polymerica, vol. 42, no. 3, pp. 8–12, 2002.
11. A. D. Orcesi and D. M. Frangopol, "Use of lifetime functions in the optimization of nondestructive inspection strategies for bridges," Journal of Structural Engineering, vol. 137, no. 4, pp. 531–539, 2011.
12. D. Cusson, Z. Lounis, and L. Daigle, "Durability monitoring for improved service life predictions of concrete bridge decks in

corrosive environments," Computer-Aided Civil and Infrastructure Engineering, vol. 26, no. 7, pp. 524–541, 2011.

13. A. K. Agrawal, A. Kawaguchi, and Z. Chen, "Deterioration rates of typical bridge elements in New York," Journal of Bridge Engineering, vol. 15, no. 4, pp. 419–429, 2010
14. G. Mitra, K. K. Jain, and B. Bhattacharjee, "Condition assessment of corrosion-distressed reinforced concrete buildings using fuzzy logic," Journal of Performance of Constructed Facilities, vol. 24, no. 6, pp. 562–570, 2010
15. A. Strauss, K. Bergmeister, S. Hoffmann, R. Pukl, and D. Novák, "Advanced life-cycle analysis of existing concrete bridges," Journal of Materials in Civil Engineering, vol. 20, no. 1, pp. 9–19, 2008.
16. G. C. Marano, G. Quaranta, and M. Mezzina, "Fuzzy time-dependent reliability analysis of RC beams subject to pitting corrosion," Journal of Materials in Civil Engineering, vol. 20, no. 9, pp. 578–587, 2008.
17. S. Tesfamariam and B. Martín-Pérez, "Bayesian belief network to assess carbonation-induced corrosion in reinforced concrete," Journal of Materials in Civil Engineering, vol. 20, no. 11, pp. 707–717, 2008.
18. J. Sobhani and A. A. Ramezaninpour, "Modeling the corrosion of reinforced concrete structures based on fuzzy systems," in Proceedings of the 3rd International Conference on Concrete and Development, pp. 729–741, Tehran, Iran, April 2009.
19. R. E. Melchers, C. Q. Li, and W. Lawanwisut, "Probabilistic modeling of structural deterioration of reinforced concrete beams under saline environment corrosion," Structural Safety, vol. 30, no. 5, pp. 447–460, 2008.
20. D. V. Val, "Deterioration of strength of RC beams due to corrosion and its influence on beam reliability," Journal of Structural Engineering, vol. 133, no. 9, pp. 1297–1306, 2007.
21. M. B. Anoop and K. Balaji Rao, "Application of fuzzy sets for remaining life assessment of corrosion affected reinforced concrete bridge girders," Journal of Performance of Constructed Facilities, vol. 21, no. 2, pp. 166–171, 2007

22. L. Shao and C. Q. Li, "Reliability based asset management strategy for concrete infrastructure,"International Journal of Materials and Structural Reliability, vol. 5, no. 1, pp. 13–28, 2007.
23. S. Sasmal, K. Ramanjaneyulu, S. Gopalakrishnan, and N. Lakshmanan, "Fuzzy logic based condition rating of existing reinforced concrete bridges," Journal of Performance of Constructed Facilities, vol. 20, no. 3, pp. 261–273, 2006.
24. C. Q. Li, W. Lawanwisut, and J. J. Zheng, "Time-dependent reliability method to assess the serviceability of corrosion-affected concrete structures," Journal of Structural Engineering, vol. 131, no. 11, pp. 1674–1680, 2005.
25. J. S. Kong and D. M. Frangopol, "Sensitivity analysis in reliability-based lifetime performance prediction using simulation," Journal of Materials in Civil Engineering, vol. 17, no. 3, pp. 296–306, 2005.
26. K. A. T. Vu and M. G. Stewart, "Predicting the likelihood and extent of reinforced concrete corrosion-induced cracking," Journal of Structural Engineering, vol. 131, no. 11, pp. 1681–1689, 2005
27. F. Akgül and D. M. Frangopol, "Lifetime performance analysis of existing reinforced concrete bridges. II: application," Journal of Infrastructure Systems, vol. 11, no. 2, pp. 129–141, 2005.
28. D. V. Val, "Effect of different limit states on life-cycle cost of RC structures in corrosive environment," Journal of Infrastructure Systems, vol. 11, no. 4, pp. 231–240, 2005
29. M. G. Stewart and D. V. Val, "Multiple limit states and expected failure costs for deteriorating reinforced concrete bridges," Journal of Bridge Engineering, vol. 8, no. 6, pp. 405–415, 2003.
30. C. Q. Li, "Life-cycle modeling of corrosion-affected concrete structures: propagation," Journal of Structural Engineering, vol. 129, no. 6, pp. 753–761, 2003
31. C. Q. Li, "Life cycle modeling of corrosion affected concrete structures-initiation," Journal of Materials in Civil Engineering, vol. 15, no. 6, pp. 594–601, 2003.
32. M.-T. Liang, L.-H. Lin, and C.-H. Liang, "Service life prediction of existing reinforced concrete bridges exposed to chloride environment," Journal of Infrastructure Systems, vol. 8, no. 3, pp. 76–85, 2002.

33. J. Gulikers, "Critical review of corrosion deterioration models for reinforced concrete," in Proceedings of the 9th Durability of Building Materials and Components (DBMC '02), pp. 0941–0910, Brisban, Australia, 2002.
34. A. Abu-Tair, C. McParland, J. F. Lyness, and A. Nadjai, "Predictive models of deterioration rates of concrete bridges using the factor method based on historic inspection data," in Proceedings of the 9th Durability of Building Materials and Components (DBMC '02), p. 262, Brisban, Australia, 2002.
35. K. Vu and M. G. Stewart, "Service life prediction of reinforced concrete structures exposed to aggressive environments," in Proceedings of the 9th Durability of Building Materials and Components (DBMC '02), p. 119, Brisban, Australia, 2002.
36. M. P. Enright and D. M. Frangopol, "Survey and evaluation of damaged concrete bridges," Journal of Bridge Engineering, vol. 5, no. 1, pp. 31–38, 2000.
37. B. Naticchia, "Quality simulation for assessment of chemical-physical compatibility among building components," Durability of Building Materials and Components, vol. 8, pp. 1457–1466, 1999.
38. L. Scheremans and D.V. Gemert, "Service prediction of reinforced concrete structures, based on in-service, chloride penetration profiles," Durability of Building Materials and Components, vol. 8, pp. 83–93, 1999.
39. B. Teply, D. Novak, Z. Kersner, and W. Lawanwisut, "Failure probability of deteriorating reinforced concrete beams," Durability of Building Materials and Components, vol. 8, pp. 1357–1366, 1999.
40. M. P. Enright and D. M. Frangopol, "Condition prediction of deteriorating concrete bridges using Bayesian updating," Journal of Structural Engineering, vol. 125. no. 10, pp. 1118–1125, 1999.
41. M. S. Cheung and B. R. Kyle, "Service life prediction of concrete structures by reliability analysis,"Construction and Building Materials, vol. 10, no. 1, pp. 45–55, 1996.
42. Y. Mori and B. R. Ellingwood, "Reliability-based service-life assessment of aging concrete structures,"Journal of Structural Engineering, vol. 119, no. 5, pp. 1600–1621, 1993.

Chapter 8

System Identification in Structural Engineering

G.F. Sirca Jr and H. Adeli

Department of Civil, Environmental, and Geodetic Engineering, The Ohio State University, 470 Hitchcock Hall, 2070 Neil Avenue, Columbus, OH, 43210, USA

ABSTRACT

A review of representative research reported in journal articles in the field of structural system identification published in journals since 1995 is presented. The paper is divided into five sections based on the general approach used: conventional model-based, biologically-inspired, signal processing-based, chaos theory, and multi-paradigm approaches. Most of the published papers deal with small and academic problems. System identification of large real-life structures with nonlinear behavior subjected to unknown dynamic loading such as strong ground motions is challenging. It is believed a multi-paradigm approach is the most effective strategy for system identification of large structures subjected to dynamic loading.

INTRODUCTION

System Identification (SI) is the process of modeling an unknown system based on a set of input–outputs and is employed in different fields of engineering [1] and [2]. In the case of structural system identification, this can be done in the form of (a) Identifying structural parameters such as stiffness, vibration signatures such as frequencies, mode shapes, and damping ratios, and stress and strain energies, or (b) Structural response. In the context of this paper SI is used for dynamic systems such as structures subjected to dynamic vehicular, seismic, wind, or impact loading.

Some SI models have been developed for identifying damage in structural elements including steel and concrete beams, suspension cables, and concrete columns. These models have applications in health monitoring of civil infrastructure such as bridges [3], [4], [5] and [6], buildings [7], [8] and [9], TV towers [10], pavements [11], and railways [12]. State-of-the-art reviews of active and semi-active control of smart structures as well as hybrid control systems have been presented recently [13] and [14].

In this work, a review of representative research, reported in articles published in journals since 1995, in the field of structural system identification, is presented. SI articles related to damage detection are limited to approaches used to identify damage in structures or structural components without any focus on sensing devices or any monitoring of structures. A subsequent complementary paper in preparation will present a review of recent articles in the area of health monitoring of structures including damage detection using an array of sensors.

This review is divided into five sections based on the general approach used:

- Conventional model-based approaches;

- Biologically-inspired approaches such as neural network and genetic algorithm;
- Signal processing-based approaches such as wavelets;
- Chaos theory;
- Multi-paradigm approaches.

Each section, except the last two is further divided into the following applications: beams and two-dimensional (2D) frames, three-dimensional (3D) building structures, bridges, and other structures. Papers on beam and 2D frame applications are mostly of academic nature. Articles on 3D building structures in general attempt to model real structures. Bridges can be modeled relatively accurately as a 2D or a 3D structure depending on its type.

CONVENTIONAL MODEL-BASED APPROACHES

Conventional model-based approaches for system identification typically use a computer model of the structure, such as a Finite-Element Method (FEM) model, to identify structural parameters primarily from field or laboratory test data [15], [16] and [17]. The advantages of using a model-based approach are modeling and estimating the physical properties and convenience. The convenience factor includes the ability to use commercially available software to create and maintain the structural model. The main disadvantage of the model-based approach is that it does not produce accurate results for large and complex real-life structures.

Beams and Two Dimensional Frames

Most beams and 2D frame structures can be modeled with reasonable accuracy using a model-based approach. Liu [18] used measured natural frequencies and mode shapes for system identification and damage detection of an aluminum 2D truss with 21 members. The damage was defined as a reduction in the axial stiffness which was determined by identifying changes to the natural frequencies and mode shapes of the truss. Kosmatka and Ricles [19] use modal vibration characterization for detecting stiffness changes in a ten-bay aluminum space truss. Araki and Miyagi [20] use a mixed integer nonlinear least-squares approach for detecting changes in member properties in a 2D bridge-type truss structure.

Because of their simplicity beams lend themselves well to the model-based approach and their dynamic loading can be modeled easily. Kim and Stubbs [21] investigate how model uncertainty affects accuracy when identifying structural degradation of a two-span aluminum plate girder, especially when only a few modal response parameters are used. The authors point out the potential shortcomings of the model-based approach especially when the model is too idealized and not a good representation of the actual structure. Modeling error-effects on model-based systems are also investigated by Sanayei et al. [22] with respect to the error functions used. The authors compare the performance of two stiffness-based and two flexibility-based error functions in terms of model error propagation rate and the quality of the final parameter estimates. They conclude that stiffness-based error functions are better than flexibility-based functions in terms of modeling error. Many other authors have applied this simple approach to mostly small beams and trusses [17], [23], [24], [25], [26] and [27].

Damage identification in beams is a common theme in system identification research. Kim and Stubbs [28] discuss damage identification of a two-span continuous beam using modal information. Lee and Shin [29] detect changes in the stiffness of beams based on a frequency-based response function. Jiang and Wang [30] use a frequency-shift-based method for damage detection of a cantilevered beam benchmark example. Lu et al. [31] use a Finite Element (FE) model updating method to detect changes in flexural stiffness in coupled beam systems (two beams connected by a set of linear and rotational springs).

2D frames also lend themselves well to the model-based system identification approach. Bodeux and Golinval [32] apply the Auto-Regressive Moving Average Vector (ARMAV) method for identifying the changes in stiffnesses of two-story, single-bay steel frames. Hung and Ko [33] apply a Vector Backward Autoregressive (VBAR) approach to identify modal properties of a simple beam. Hung et al. [34] extend the VBAR method to include exogenous data and apply it to a half-scale physical model of a five-story, single-bay steel structure.

Peterson et al. [15] and [16] determine localized damage in timber beams based on the comparison of the differences in modal strain energy between undamaged and damaged structures. Ndambi et al. [35] use eigenfrequencies and mode shape derivatives for detecting cracks in Reinforced Concrete (RC) beams. Ren and De Roeck [36] and [37] and Unger et al. [38] use modal data to find the location of damage and the severity of the damage defined as a change in the stiffness of concrete beams and prestressed concrete beams, respectively. Pavlenko and Loh [39] use pseudodynamic testing data of a full-scale three-story RC moment frame for nonlinear system identification. Robert-Nicoud et al. [40] use model decomposition and stochastic search to identify the material properties of a timber beam. Moaveni et al. [41] use a FE updating method to determine changes in the stiffness of beams and moment-resisting frames subjected to seismic loading, and a 2/3-scale three-story masonry-infilled RC frame structure tested on a shake table, respectively.

Huang et al. [42] present a time-varying autoregressive approach with exogenous input approach for identifying the modal parameters of a frame subjected to a series of base excitations on a shaking table test. Adewuyi and Wu [43] propose damage locating indices based on normalized modal macrostrain vectors to locate damage in beam-like structures.

Building Structures and three Dimensional Frames

Ghanem and Shinozuka [44] and [45] provide a review of methods of system identification by application to experimental data obtained on three- and five-story steel building structures subjected to seismic loading, including the extended Kalman filter, maximum-likelihood technique, recursive least-squares, and recursive instrumental-variable method. Koh et al. [46] determine the stiffness changes in a 12-story building with varying degrees of noisy data and a 6-story laboratory model subjected to hammer tests. Zhu and Xu [47], Fritzen and Bohle [48], and Görl and Link [49] perform similar work on a 3-story steel moment space frame, a 2-story steel frame deformed by mechanical pistons, and a 2-story full-scale steel frame subjected to seismic loading, respectively. Kunnath [50] also uses the modal identification method for seismic evaluation and design of moment resisting frame buildings. Haralampidis et al. [51] use pushover analysis for system identification of scaled 8- and 16-story 3-bay by 5-bay concrete frame space frames. Liu et al. [52] examine ways to calibrate the FE model used for system identification of a 14-story five-bay steel moment frame office building constructed in the early 1980's subject to seismic loading. Zhao et al. [53] use a least-squares method for modal identification of a seismically-loaded 3-story shear frame.

Moaveni et al. [54] examine six variations of model-based approach including, data-driven stochastic subspace identification, frequency domain decomposition, observer/Kalman filter identification, and general realization algorithm for system identification of a full-scale 7-story RC building structure subjected to shake table loading, and assert that probabilistic system identification methods in connection with FE model updating provide the most desirable results. Saito and Beck [55] present a Bayesian framework [56] and [57]for model order selection in auto-regressive exogenous models for system identification and health monitoring of a high-rise building. They apply the framework to one of the twin 24-story steel-framed buildings located in Tokyo, Japan. Kim and Lynch [58] use subspace system identification on a scaled model of a 6-story steel frame structure subjected to shake table loading.

Figueiredo et al. [59] compare three different autoregressive (AR) methods to determine the influence of the model order on the damage detection process using the test results on a three-story base-excited frame structure.

Bridges

Bridges have specific issues that do not exist in building-type structures such as vehicle or pedestrian loading in addition to wind and seismic loading. Vehicle loading adds to the difficulty in accurate modeling of a bridge with respect to the fatigue due to repeated and long-term loading. Aktan et al. [60] discuss system identification of a steel girder highway bridge using FE model updating. Aktan et al. [61] continue the research by considering two-dimensional grid models of the bridge deck subjected to vehicle loading. Alaylioglu and Alaylioglu [62] use finite-element model updating with *in situ* testing for identifying stiffness changes of a highway bridge subjected to vehicle loading. Huang et al. [63] identify the vibration frequencies, mode shapes, and damping rations of a three-span box-girder concrete highway bridge using the results of free vibration tests on the bridge. Macdonald and Daniell [64] perform a similar study on a cable-stayed bridge. Gu and Qin [65] identify flutter and aerodynamic effects of bridge decks of a long-span cable-stayed bridge with a cross section modeled after the Hong-Guang Bridge in China subjected to wind loading using a covariance-driven stochastic subspace identification technique. Robert-Nicoud et al. [66] employ an FE model-updating method and field measurements of deflections, rotations, and strain for identification of the 395 m three-span prestressed concrete box girder Lutrive Bridge in Switzerland. Chen et al. [67] use an autoregressive-moving-average method for identification of a 3-span prestressed concrete box girder bridge under traffic excitations. González et al. [68] present an approach for monitoring and identifying damping of three-span highway bridges using a vehicle fitted with accelerometers driving over the bridge. Pridham and Wilson [69] evaluate the dynamic characteristics of the Quincy Bayview cable-stayed bridge in Quincy, IL using output-only system identification. Altuni ik et al. [70] use an output-only model-based approach for system identification of a post-tensioned segmental twin box girder three-span concrete highway bridge subjected to ambient vibrations.

He et al. [71] apply three different system identification methods, two model-based approaches using eigenvalues and the state space approach and a frequency domain decomposition method, to the Alfred Zampa Memorial suspension bridge near San Francisco subjected to ambient and forced vibrations. Chaudhary et al. [72] present system identification of two base-isolated bridges subjected to seismic loading using FE updating. Siringoringo and Fujino [73] compare FE model updating with two time-domain SI methods including the natural excitation technique combined with the eigensystem realization algorithm for identification of three cable-stayed bridges, the Yokohama Bay Bridge, Rainbow Bridge, and Tsurumi Fairway Bridge in Japan subjected to ambient vibrations and conclude that the time-domain method *"is more practical and efficient especially when applied to voluminous data from multi-channel measurement"*.

Nagayama et al. [74] apply an inverse analysis model-based method for identifying the structural properties of the three-span Hakucho suspension bridge in Japan subjected to ambient vibration. Hong et al. [75] use FE model updating to identify the response of the New Carquinez suspension bridge in Northern California subject to wind loading. The bridge is fitted with wind and seismic monitoring sensors. Tan and Huang [76]present system identification of a base-isolated highway bridge with lead-rubber bearings using a bilinear hysteretic model. Luco and de Barros [77] consider structure-soil interaction for system identification of a $\frac{1}{4}$-scale model of Hualien RC bridge in Taiwan.

Damage identification of bridges is the focus in some research. Bolton et al. [78] use modal testing of a concrete box-girder bridge to develop a nondestructive damage-evaluation process that determines the rate of structural deterioration and remaining service life. Jang et al. [79] use FE model updating for identifying cracks in a single-span steel girder bridge. Peterson et al. [80] use FE model updating to identify decay in a simple three-girder timber bridge tested in a laboratory.

Gangone et al. [81] describe deployment of a wireless sensor system for identification of the modal characteristics of a three span simply supported bridge in the state of New York including natural frequencies and mode shapes.

Talebinejad et al. [82] study the variability of the enhanced coordinate modal assurance criterion, damage index method, mode shape curvature method, and modal flexibility index method to identify mode shapes and natural frequencies of undamaged Bayview Bridge, a cable-stayed bridge in Quincy, Illinois, and the same structure simulated with damage. They conclude that *"Damage index and mode shape curvature perform well even when a few sensors are used."*

Other Structures

Law et al. [83] use FE model updating for the identification of wind loading on a guyed mast. Alves and Hall [84] use FE model updating for identification of a concrete arch dam subjected to seismic loading taking into account its interaction with the foundation rock.

NEURAL NETWORKS AND GENETIC ALGORITHMS

Model-based system identification methods cannot be used effectively for large and complicated real-world structures with nonlinear behavior. For such cases, biologically-inspired or soft computing techniques such as Neural Networks (NN) [85], [86], [87], [88], [89], [90], [91] and [92], Genetic Algorithms (GA) [93], [94],[95], [96], [97] and [98], or particle swarm optimization [99] have been proposed as a more effective approach.

The accuracy of the NN models depends on how it is trained to solve new problems. A poorly trained model using sparse and/or corrupted data leads to inaccurate results. Pierce et al. [100] address the reliability of the NN models for damage detection when network training data are incomplete and how to improve the model. Some researchers have used biologically-inspired methods to enhance model-based methods in an attempt to reduce the shortcomings of the traditional model-based approach.

Beams and Two-Dimensional Frames

Franco et al. [101] use an evolutionary algorithm to identify the structural parameters of a 10-DOF shear frame. Raich and Liszhai [102] use GA to identify the stiffness changes in a steel beam and a 3-story, 3 bay frame. Zhao et al. [103] use a Counter-Propagation Neural network (CPN) [104] and [105] to detect stiffness changes in a 3-span beam and plane frame. Bani-Hani et al. [106] use experimental data from a three-story steel frame structure subjected to simulated earthquake forces to train a multilayer feedforward NN for structural identification and control. Backpropagation neural network (BPN) [107] is used by Zang and Imregun [108] to detect fatigue cracks in steel beams, by Hung and Kao [109] to identify changes in stiffness and damping of a 5-story shear building structure, by Ni et al. [110] to identify connection failures in steel frames, and by Huang et al. [111] to identify the dynamic characteristics of the scaled model of a five-story steel frame subjected to earthquake loading on a shake table.

Hao and Xia [112] use a GA to identify a reduction in the cross sectional area, which affects the stiffness and flexibility of the structure of a cantilevered beam and a portal frame. Perry et al. [113] use a GA for output-only structural identification of a 20-DOF 2D frame.

Building and Three-Dimensional Frame Structures

Yen [114] uses a radial basis function NN [115], [116], [117], [118], [119], [120] and [121] for the identification and control of large structures. Masri et al. [122] use a feed-forward NN [123] with two hidden layers for the detection of changes in structural parameters. They use vibration measurements from a*healthy* structure to train the NN to identify damage to the structure, defined as changes in the stiffness and damping values. Zapico et al. [124] apply a multi-layer perceptron NN for damage detection of steel frames where the network is trained using results from FE simulations. BPN is used by Vaidyanathan et al. [125] to identify the response of building structures fitted with viscoelastic dampers who also determine the number of dampers needed to reduce the peak displacement for new building designs and for retrofitting existing buildings.

Na et al. [126] use a GA for identifying stiffness changes in a 20-story model of a shear frame. Marano et al. [127] also use a GA for system identification of a shear frame with incomplete measurements.

Bridges

Pandey and Barai [128] use multilayer perceptron and BPN to determine the stiffness reduction of steel truss bridges. Huang and Loh [129] use a multi-layer Levenberg–Marquardt BPN for system identification of a 5-span prestressed concrete box girder bridge subjected to seismic loading. BPN is used by Wu et al. [130] in a decentralized parametric method for identification of stiffness changes in bridge structures, by Chen [131]for identification of the modal parameters of a cable-stayed bridge and a three-span highway bridge, and by Xu and Humar [132] for identification of stiffness changes in a girder bridge. Both Wu et al. [130] and Xu and Humar [132] apply the method to the Crowchild 3-span steel girder bridge in Calgary, Canada.

Jafarkhani and Masri [133] evaluate the performance of an evolutionary strategy in the FE updating approach for damage detection in a quarter-scale two-span RC bridge. Mosquera et al. [134] use a GA to identify the changes in the displacement of a two-span RC box girder highway bridge in El Centro, CA.

Other Structures

Kang et al. [135] propose a biologically-inspired method based on combining the Nelder–Mead simplex method with artificial bee colony algorithm, a particle swarm optimization method [99] for system identification of a concrete gravity dam and a concrete arc dam.

WAVELETS AND OTHER SIGNAL PROCESSING APPROACHES

In the past two decades because of their ability to retain both time and frequency information, wavelets have been used increasingly to solve complicated time series pattern recognition problems in different areas such as transportation engineering [136], [137], [138], [139], [140] and [141] in addition to structural engineering [142], [143], [144], [145] and [146]. Wavelets are also used as part of a multi-paradigm system identification approach to be described in the next section.

Beams

Liew and Wang [147] use wavelets to identify cracks in simply supported beams. Pan and Lee [148] use wavelets to identify yielding of lumped mass Single-Degree-Of-Freedom (SDOF) and Multi-Degree-Of-Freedom (MDOF) systems subjected to seismic loading. Bao et al. [149] employ the Hilbert-Huang transform for system identification of concrete-steel composite beams.

Two-Dimensional Frames

Kijewski and Kareem [150] compare wavelets with Fourier transform for system identification of a tower structure subjected to a typhoon and conclude that the wavelet transform is superior to the Fourier transform due to its widely-recognized multi-resolution, time-frequency analysis capabilities. Qiao et al. [151] also compare wavelets with the Fast Fourier Transform (FFT) for identifying stiffness changes in a 2D 3-story steel frame and conclude that wavelets result in higher pattern-matching resolution than the FFT. Huang et al. [152] use wavelets for structural modal parameter identification of steel frames subjected to seismic loading.

Tee et al. [153] use Kalman filters for structural system identification of an 8-story steel frame structure. Banerjee et al. [154] also apply Kalman filters for system identification of a four-span truss structure model.

Building and Three-Dimensional Frame Structures

Sepe et al. [155] use Fourier transform for system identification of a multi-story frame structure where the input data is unknown. Yuen and Katafygiotis [156] use FFT for the system identification of a 10-story shear frame. Loh et al. [157] use wavelets for system identification of RC frames subjected to shake table testing. Hazra et al. [158] use wavelets for system identification of an airport control tower near Toronto, Canada. The tower is 49 m tall and consists of a ten-level steel frame with composite concrete floors at each level.

Hwang et al. [159] use Kalman filters for system identification of a five-story building subjected to wind loading. Kijewski-Correa et al. [160] present a framework and their initial efforts for structural system identification and health monitoring of the 828 m tall Burj Khalifa building, the tallest building in the world. The idea is to use data streams from distributed heterogeneous sensors installed in the building to study variations of modal participation in time using wavelet analysis.

Bridges

Lee et al. [161] apply the Hilbert transform to identify dynamic behavior of the three span, prestressed concrete box girder Bai-Ho bridge in Taiwan subjected to seismic and vehicle loading. Huang et al. [152] use wavelets for structural modal parameter identification of a five-span arch bridge subjected to seismic loading. Yan and Miyamoto [162] present a comparative study of modal parameter identification of a benchmark bridge structure using the wavelet and the Hilbert-Huang transforms and conclude that the former performs slightly better than the latter.

Other Structures

Wavelets are used by Douka et al. [163], Kim et al. [164], and Rucka and Wilde [165] to detect damage in plates, by Castro et al. [166] to detect damage in rods, by Le and Argoul [167], and Chakraborty et al. [168]and Chen et al. [169] for modal identification of general mass-damper systems, and Ching and Glaser [170]for tracking changes to soil slopes subjected to seismic loading.

Chaos Theory

A few researchers have employed the chaos theory and the fractal concept [171] to model the complicated structural dynamics for system identification. Schoefs et al. [172] use polynomial chaos representation for identification of mechanical characteristics of complex instrumented structures such as harbor structures. Li et al. [6] present a method for estimation of fatigue damage in Fiber Reinforced Polymer (FRP) stay cables using acoustic emission technique and the fractal concept from the chaos theory. Li et al. [173] describe a method based on Katz's Fractal Dimension (FD) measure of displacement mode shape for damage localization in beams.

MULTI-PARADIGM APPROACHES

These approaches integrate two or more computing paradigms such as NNs [174], Fuzzy Logic (FL) [175],[176], [177] and [178], evolutionary computing/GA [179], signal processing techniques such as wavelets, and the chaos theory to come up with a more powerful approach especially for nonlinear and complex problems. The multi-paradigm approach to problem solving has attracted the attention of researchers since Adeli and Hung [174] (1995) advocated the integration of the three principal fields of computational intelligence, NN, FL, and GA, for solution of complicated pattern recognition problems such as face recognition and engineering design. Development of FL–NN algorithms has been particularly popular [180],[181], [182], [183], [184] and [185] followed by FL–GA [186] and NN–GA [187] algorithms.

Wadia-Fascetti and Gunes [188] combine FL and statistical analysis to develop a method for creating earthquake response spectra for dynamic analysis of building structures. Adeli and Jiang [189] present a multi-paradigm dynamic time-delay fuzzy Wavelet Neural Network (WNN) model for nonparametric identification of structures with nonlinear behavior using the nonlinear autoregressive moving average with exogenous inputs. The model is based on the integration of four different computing concepts: dynamic time delay NN, wavelet, FL, and the reconstructed state space concept from the chaos theory. Noise in the signals is removed using the discrete wavelet packet transform method. In order to preserve the dynamics of time series, the reconstructed state space concept from the chaos theory is employed to construct the input vector. In addition to denoising, wavelets are employed in combination with NNs and FL to create a new pattern recognition model to capture the characteristics of the time series sensor data accurately and efficiently. The model balances the global and local influences of the training data and incorporates the imprecision existing in the sensor data effectively. Experimental results of a five-story steel frame are employed to validate the computational model and demonstrate its accuracy and efficiency. Compared with conventional NNs, training of a dynamic NN for system identification of large-scale structures is substantially more complicated and time-consuming because both input and output of the network are not single-valued but thousands of time steps. Jiang and Adeli [190] (2005) present an adaptive Levenberg–Marquardt-least squares algorithm with a backtracking inexact linear search scheme for training of the dynamic fuzzy WNN model. The approach avoids the second-order differentiation required in the Gauss–Newton algorithm and overcomes the numerical instabilities encountered in the steepest descent algorithm with an improved learning convergence rate and high computational efficiency. The model is applied to two highrise moment-resisting building structures taking into account their geometric nonlinearities. Validation results demonstrate that the new methodology provides an efficient and accurate tool for nonlinear system identification of high-rising buildings.

Jiang and Adeli [7] present a nonparametric system identification-based model for damage detection of irregular highrise building structures subjected to seismic excitations using the dynamic fuzzy WNN model with an adaptive LM–LS learning algorithm. A new damage evaluation method is proposed based on a power density spectrum method. A multiple signal classification (MUSIC) method is developed to compute the pseudospectrum from the structural response time series. The methodology is validated using the data obtained for a 38-story concrete test model.

Recently, Osornio-Rios et al. [191] verified the MUSIC algorithm introduced first by Jiang and Adeli [7] for health monitoring of structures based on an extensive experimental study using a five-bay truss-type structure, and presented a methodology for identifying, locating, and and quantifying the severity of corrosion and crack damage.

Final Remarks

A review of journal articles published on system identification of structures was presented in this paper. This remains an active area of research because it has direct applications in the increasingly important subject of health monitoring of structures. Most of the published papers deal with small and academic problems. System identification of large real-life structures with nonlinear behavior subjected to unknown dynamic loading such as strong ground motions is challenging. It is believed a multi-paradigm approach is the most effective for modeling and solution of this complicated problem.

REFERENCES

1. Vejmelka, M., Palus, M. and Susmakova, K. "Identification of nonlinear oscillatory activity embedded in broadband neural signals", International Journal of Neural Systems, 20(2), pp. 117–128 (2010).
2. Puscasu, G. and Codres, B. "Nonlinear system identification and control based on modular neural networks", International Journal of Neural Systems, 21(4), pp. 319–334 (2011).

3. Chen, B. and Liu, W. "Mobile agent computing paradigm for building a flexible structural health monitoring sensor network", Computer-Aided Civil and Infrastructure Engineering, 25(7), pp. 504–516 (2010).
4. Bocca, M., Mahmood, A., Eriksson, L.M., Kullaa, J. and Jäntti, R. "A synchronized wireless sensor network for experimental modal analysis in structural health monitoring", Computer-Aided Civil and Infrastructure Engineering, 26(7), pp. 483–499 (2011).
5. Cusson, D., Lounis, Z. and Daigle, L. "Monitoring life cycle performance of aging concrete highway bridges built in corrosive environments", Computer-Aided Civil and Infrastructure Engineering, 26(7), pp. 524–541 (2011).
6. Li, H., Huang, Y., Chen, W.L., Ma, M.L., Tao, D.W. and Ou, J.P. "Estimation and warning of fatigue damage of FRP stay cables based on acoustic emission technique and fractal theory", Computer-Aided Civil and Infrastructure Engineering, 26(7), pp. 500–512 (2011).
7. Jiang, X. and Adeli, H. "Pseudospectra, MUSIC, and dynamic wavelet neural network for damage detection of highrise buildings", International Journal for Numerical Methods in Engineering, 71(5), pp. 606–629 (2007).
8. Park, H.S., Lee, H.M., Adeli, H. and Lee, I. "A new approach for health monitoring of structures: terrestrial laser scanning", Computer-Aided Civil and Infrastructure Engineering, 22(1), pp. 19–30 (2007).
9. Lee, H.M. and Park, H.S. "Gauge-free stress estimation of a beam-like structure based on terrestrial laser scanning", Computer-Aided Civil and Infrastructure Engineering, 26(8), pp. 647–658 (2011).
10. Xia, Y., Ni, Y.Q., Zhang, P., Liao, W.Y. and Ko, J.M. "Stress development of a super-tall structure during construction: numerical analysis and field monitoring verification", Computer-Aided Civil and Infrastructure Engineering, 26(7), pp. 542–559 (2011).
11. Lajnef, N., Rhimi, M., Mhamdi, L., Chatti, K. and Faridazar, F. "Toward a smart, long-term pavement monitoring system", Computer-Aided Civil and Infrastructure Engineering, 26(7), pp. 513–523 (2011).

12. Stratman, B., Mahadevan, S., Li, C. and Biswas, G. "Identification of critical inspection samples among railroad wheels by similarity-based agglomerative clustering", Integrated Computer-Aided Engineering, 18(3), pp. 203–219 (2011).
13. Fisco, N.R. and Adeli, H. "Smart structures: part I — active and semiactive control", Scientia Iranica — Transactions A: Civil Engineering, 18(3), pp. 275–284 (2011).
14. [14] Fisco, N.R. and Adeli, H. "Smart structures: part II — hybrid control systems and control strategies", Scientia Iranica — Transactions A: Civil Engineering, 18(3), pp. 285–295 (2011).
15. Peterson, S., McLean, D., Symans, M., Pollock, D., Cofer, W., Emerson, R. and Fridley, K. "Application of dynamic system identification to timber beams. I", Journal of Structural Engineering, 127(4), pp. 418–425 (2001).
16. Peterson, S., McLean, D., Symans, M., Pollock, D., Cofer, W., Emerson, R. and Fridley, K. "Application of dynamic system identification to timber beams. II", Journal of Structural Engineering, 127(4), pp. 426–432 (2001).
17. Liu, P. and Chian, C. "Parametric identification of truss structures using static strains", Journal of Structural Engineering, 123(7), pp. 927–933 (1997).
18. Liu, P. "Identification and damage detection of trusses using modal data", Journal of Structural Engineering, 121(4), pp. 599–608 (1995).
19. Kosmatka, J. and Ricles, M. "Damage detection in structures by modal vibration characterization", Journal of Structural Engineering, 125(12), pp. 1384–1392 (1999).
20. Araki, Y. and Miyagi, Y. "Mixed integer nonlinear least-squares problem for damage detection in truss structures", Journal of Engineering Mechanics, 131(7), pp. 659–667 (2005).
21. Kim, J. and Stubbs, N. "Model-uncertainty impact and damage-detection accuracy in plate girder", Journal of Structural Engineering, 121(10), pp. 1409–1417 (1995).
22. Sanayei, M., Wadia-Fascetti, S., Arya, B. and Santini, E. "Significance of modeling error in structural parameter estimation", Computer-Aided Civil and Infrastructure Engineering, 16(1), pp. 12–27 (2001).

23. Sheinman, I. "Damage detection and updating of stiffness and mass matrices using mode data", Computers & Structures, 59(1), pp. 149–156 (1996).
24. Shi, Z., Law, S. and Zhang, L. "Damage localization by directly using incomplete mode shapes", Journal of Engineering Mechanics, 126(6), pp. 656–660 (2000).
25. Pothisiri, T. and Hjelmstad, K. "Structural damage detection and assessment from modal response", Journal of Engineering Mechanics, 129(2), pp. 135–145 (2003).
26. Kim, J., Ryu, Y., Cho, H. and Stubbs, N. "Damage identification in beam-type structures: frequency-based method vs. mode-shape-based method", Engineering Structures, 25(1), pp. 57–67 (2003).
27. Ge, M. and Lui, E. "Structural damage identification using system dynamic properties", Computers & Structures, 83(27), pp. 2185–2196 (2005).
28. Kim, J. and Stubbs, N. "Improved damage identification method based on modal information", Journal of Sound and Vibration, 252(2), pp. 223–238 (2002).
29. Lee, U. and Shin, J. "A frequency response function-based structural damage identification method", Computers & Structures, 80(2), pp. 117–132 (2002).
30. Jiang, L. and Wang, K. "An enhanced frequency-shift-based damage identification method using tunable piezoelectric transducer circuitry", Smart Materials and Structures, 15(3), pp. 799–808 (2006).
31. Lu, Z.R., Liu, J.K., Huang, M. and Xu, W.H. "Identification of local damages in coupled beam systems from measured dynamic responses", Journal of Sound and Vibration, 326(2), pp. 177–189 (2009).
32. Bodeux, J. and Golinval, J. "Application of ARMAV models to the identification and damage detection of mechanical and civil engineering structures", Smart Materials and Structures, 10(3), pp. 479–489 (2001).
33. Hung, C. and Ko, W. "Identification of modal parameters from measured output data using a vector backward autoregressive model", Journal of Sound and Vibration, 256(2), pp. 249–270 (2002).

34. Hung, C., Ko, W. and Peng, Y. "Identification of modal parameters from measured input and output data using a vector backward autoregressive with exogenous model", Journal of Sound and Vibration, 276(3–5), pp. 1043–1063 (2004).
35. Ndambi, J., Vantomme, J. and Harri, K. "Damage assessment in reinforced concrete beams using eigenfrequencies and mode shape derivatives", Engineering Structures, 24(4), pp. 501–515 (2002).
36. Ren, W. and De Roeck, G. "Structural damage identification using modal data. I: simulation verification", Journal of Structural Engineering, 128(1), pp. 87–95 (2002).
37. Ren, W. and De Roeck, G. "Structural damage identification using modal data. II: test verification", Journal of Structural Engineering, 128(1), pp. 96–104 (2002).
38. Unger, J., Teughels, A. and De Roeck, G. "Damage detection of a prestressed concrete beam using model strains", Journal of Structural Engineering, 131(9), pp. 1456–1463 (2005).
39. Pavlenko, O. and Loh, C. "Nonlinear system identification of reinforced concrete steel structure: using pseudodynamic testing data", Journal of Engineering Mechanics, 130(7), pp. 836–847 (2004).
40. Robert-Nicoud, Y., Raphael, B. and Smith, I. "System identification through model composition and stochastic search", Journal of Computing in Civil Engineering, 19(3), pp. 239–247 (2005).
41. Moaveni, B., Stravridis, A., Lombaert, G., Conte, J. and Shing, P. "Finite element model updating for assessment of progressive damage in a three-story infilled RC frame", Journal of Structural Engineering, (2012), http://dx.doi.org/10.1061/(ASCE)ST.1943-541X.0000586.
42. Huang, C.S., Huang, S.L., Su, W.C. and Wu, C.L. "Identification of timevariant modal parameters using TVARX and low-order polynomial function", Computer-Aided Civil and Infrastructure Engineering, 24(7), pp. 470–491 (2009).
43. Adewuyi, A.P. and Wu, Z. "Vibration-based damage localization in flexural structures using normalized modal macrostrain techniques from limited measurements", Computer-Aided Civil and Infrastructure Engineering, 26(3), pp. 154–172 (2011).

44. Ghanem, R. and Shinozuka, M. "Structural-system identification. I: theory", Journal of Engineering Mechanics, 121(2), pp. 255–264 (1995).

45. Ghanem, R. and Shinozuka, M. "Structural-system identification II: experimental verification", Journal of Engineering Mechanics, 121(2), pp. 265–273 (1995).

46. Koh, C., See, L. and Balendra, T. "damage detection of buildings: numerical and experimental studies", Journal of Structural Engineering, 121(8), pp. 1155–1160 (1995).

47. Zhu, H. and Xu, Y. "Damage detection of mono-coupled periodic structures based on sensitivity analysis of modal parameters", Journal of Sound and Vibration, 285(1–2), pp. 365–390 (2005).

48. Fritzen, C. and Bohle, K. "Application of model-based damage identification to a seismically loaded structure", Smart Materials and Structures, 10(3), pp. 452–458 (2001).

49. Görl, E. and Link, M. "Identification of damage parameters of a fullscale steel structure damaged by seismic loading", Smart Materials and Structures, 10(3), pp. 459–465 (2001).

50. Kunnath, S. "Identification of modal combinations for nonlinear static analysis of building structures", Computer-Aided Civil and Infrastructure Engineering, 19(4), pp. 246–259 (2004).

51. Haralampidis, Y., Papadimitriou, C. and Pavlidou, M. "Multi-objective framework for structural model identification", Earthquake Engineering and Structural Dynamics, 34(6), pp. 665–685 (2005).

52. Liu, H., Yang, Z. and Gaulke, M. "Structural identification and finite element modeling of a 14-story office building using recorded data", Engineering Structures, 27(3), pp. 463–473 (2005).

53. Zhao, X., Xu, Y., Li, J. and Chen, J. "Hybrid identification method for multistory buildings with unknown ground motion: theory", Journal of Sound and Vibration, 291(1–2), pp. 215–239 (2006).

54. Moaveni, B., He, X., Conte, J., Restrepo, J. and Panagiotou, M. "System identification of a 7-story full-scale building slice tested on the UCSDNEES shake table", Journal of Structural Engineering, 137(6), pp. 705–717 (2011).

55. Saito, T. and Beck, J. "Bayesian model selection for ARX models and its applications to structural health monitoring", Earthquake Engineering and Structural Dynamics, 39(15), pp. 1737–1759 (2010).

56. Cheung, S.H. and Beck, J.L. "Calculation of the posterior probability for Bayesian model class selection and averaging from posterior samples based on dynamic system data", Computer-Aided Civil and Infrastructure Engineering, 25(5), pp. 304–321 (2010).

57. Yin, T., Lam, H.F. and Chow, H.M. "A Bayesian probabilistic method for crack characterization in plate structures", Computer-Aided Civil and Infrastructure Engineering, 25(5), pp. 375–386 (2010).

58. Kim, J. and Lynch, J. "Subspace system identification of supportexcited structures – part I: theory and black-box system identification", Earthquake Engineering and Structural Dynamics, (2012), http://dx.doi.org/10.1002/eqe.2184.

59. Figueiredo, E., Park, G., Figueiras, J., Farrar, C. and Worden, K. "Influence of the autoregressive model order on damage detection", Computer-Aided Civil and Infrastructure Engineering, 26(3), pp. 225–238 (2011).

60. Aktan, A., Farhey, D., Helmicki, A., Brown, D., Hunt, V., Lee, K. and Levi, A. "Structural identification for condition assessment: experimental arts", Journal of Structural Engineering, 123(12), pp. 1674–1684 (1997).

61. Aktan, E., Catbas, N., Türer, A. and Zhang, Z. "Structural identification: analytical aspects", Journal of Structural Engineering, 124(7), pp. 817–829 (1998).

62. Alaylioglu, H. and Alaylioglu, A. "Dynamic structural assessment of a highway bridge via hybrid FE model and in situ testing", Computers & Structures, 63(3), pp. 439–453 (1997).

63. Huang, C.S., Yang, Y., Lu, L. and Chen, C. "Dynamic testing and system identification of a multi-span highway bridge", Earthquake Engineering and Structural Dynamics, 28(8), pp. 857–878 (1999).

64. Macdonald, J. and Daniell, W. "Variation of modal parameters of a cablestayed bridge identified from ambient vibration measurements and FE modelling", Engineering Structures, 27(12), pp. 1916–1930 (2005).

65. Gu, M. and Qin, X. "Direct identification of flutter derivatives and aerodynamic admittances of bridge decks", Engineering Structures, 26(14), pp. 2161–2172 (2004).

66. Robert-Nicoud, Y., Raphael, B., Burdet, O. and Smith, I. "Model identification of bridges using measurement data", Computer-Aided Civil and Infrastructure Engineering, 20(2), pp. 118–131 (2005).

67. Chen, Y., Feng, M. and Tan, C. "Modelling of traffic excitation for system identification of bridge structures", Computer-Aided Civil and Infrastructure Engineering, 21(1), pp. 57–66 (2006).

68. González, A., Obrien, E.J. and McGetrick, P.J. "Identification of damping in a bridge using a moving instrumented vehicle", Journal of Sound and Vibration, 331(18), pp. 4115–4131 (2012).

69. Pridham, B. and Wilson, J. "A reassessment of dynamic characteristics of the Quincy Bayview bridge using output-only identification techniques", Earthquake Engineering and Structural Dynamics, 34(7), pp. 787–805 (2005).

70. Altunişik, A., Bayraktar, A. and Sevim, B. "Output-only identification of posttensioned segmental concrete highway bridges", Journal of Bridge Engineering, 16(2), pp. 259–266 (2011).

71. He, X., Moaveni, B., Conte, J., Elgamal, A. and Masri, S. "System identification of Alfred Zampa memorial bridge using dynamic field test data", Journal of Structural Engineering, 135(1), pp. 54–66 (2009).

72. Chaudhary, M., Abe, M., Fujino, Y. and Yoshida, J. "System identification of two base-isolated bridges using seismic records", Journal of Structural Engineering, 126(10), pp. 1187–1195 (2000).

73. Siringoringo, D.M. and Fujino, Y. "System identification of a suspension bridge from ambient vibration response", Engineering Structures, 30(2), pp. 462–477 (2008).

74. Nagayama, T., Abe, M., Fujino, Y. and Ikeda, K. "Structural identification of a nonproportionally damped system and its application to a fullscale suspension bridge", Journal of Structural Engineering, 131(10), pp. 1536–1545 (2005).

75. Hong, A., Ubertini, F. and Betti, R. "Wind analysis of a suspension bridge: identification and finite-element model simulation", Journal of Structural Engineering, 137(1), pp. 133–142 (2011).
76. Tan, R. and Huang, M. "System identification of a bridge with lead-rubber bearings", Computers & Structures, 74(3), pp. 267–280 (2000).
77. Luco, J. and de Barros, F. "Identification of structural and soil properties from vibration tests of the Hualien containment model", Earthquake Engineering and Structural Dynamics, 34(1), pp. 21–48 (2005).
78. Bolton, R., Stubbs, N., Choi, S. and Sikorsky, C. "Documentation of changes in modal properties of a concrete box-girder bridge due to environmental and internal conditions", Computer-Aided Civil and Infrastructure Engineering, 16(1), pp. 42–57 (2001).
79. Jang, J., Yeo, I., Shin, S. and Chang, S. "Experimental investigation of system-identification-based damage assessment on structures", Journal of Structural Engineering, 128(5), pp. 673–682 (2002).
80. Peterson, S., McLean, D. and Pollock, D. "Application of dynamic system identification to timber bridges", Journal of Structural Engineering, 129(1), pp. 116–124 (2003).
81. Gangone, M.V., Whelan, M.J. and Janoyan, K.D. "Wireless monitoring of a multi-span bridge superstructure for diagnostic load testing and system identification", Computer-Aided Civil and Infrastructure Engineering, 26(7), pp. 560–579 (2011).
82. Talebinejad, I., Fischer, C. and Ansari, F. "Numerical evaluation of vibration based methods for damage assessment of cable stayed bridges", Computer-Aided Civil and Infrastructure Engineering, 26(3), pp. 239–251 (2011).
83. Law, S., Bu, J. and Zhu, X. "Time-varying wind load identification from structural responses", Engineering Structures, 27(10), pp. 1586–1598 (2005).
84. Alves, S. and Hall, J. "System identification of a concrete arch dam and calibration of its finite element model", Earthquake Engineering and Structural Dynamics, 35(11), pp. 1321–1337 (2006).

85. Adeli, H. and Park, H.S. "A neural dynamics model for structural optimization — theory", Computers & Structures, 57(3), pp. 383–390 (1995).
86. Adeli, H. and Karim, A. "Neural network model for optimization of cold-formed steel beams", Journal of Structural Engineering, 123(11), pp. 1535–1543 (1997).
87. Park, H.S. and Adeli, H. "Distributed neural dynamics algorithms for optimization of large steel structures", Journal of Structural Engineering, 123(7), pp. 880–888 (1997).
88. Samant, A. and Adeli, H. "Feature extraction for traffic incident detection using wavelet transform and linear discriminant analysis", ComputerAided Civil and Infrastructure Engineering, 15(4), pp. 241–250 (2000).
89. Graf, W., Freitag, S., Kaliske, M. and Sickert, J.U. "Recurrent neural networks for uncertain time-dependent structural behavior", ComputerAided Civil and Infrastructure Engineering, 25(5), pp. 322–333 (2010).
90. Reuter, U. and Moeller, B. "Artificial neural networks for forecasting of fuzzy time series", Computer-Aided Civil and Infrastructure Engineering, 25(5), pp. 363–374 (2010).
91. Ahmadlou, M. and Adeli, H. "Enhanced probabilistic neural network with local decision circles: a robust classifier", Integrated Computer-Aided Engineering, 17(3), pp. 197–210 (2010).
92. Freitag, S., Graf, W. and Kaliske, M. "Recurrent neural networks for fuzzy data", Integrated Computer-Aided Engineering, 18(3), pp. 265–280 (2011).
93. Kim, H. and Adeli, H. "Discrete cost optimization of composite floors using a floating point genetic algorithm", Engineering Optimization, 33(4), pp. 485–501 (2001).
94. Lee, Y. and Wei, C.H. "A computerized feature selection using genetic algorithms to forecast freeway accident duration times", Computer-Aided Civil and Infrastructure Engineering, 25(2), pp. 132–148 (2010).
95. Martínez-Ballesteros, M., Troncoso, A., Martínez-Álvarez, F. and Riquelm, J.C. "Mining quantitative association rules based on evolutionary computation and its application to atmospheric pollution", Integrated Computer-Aided Engineering, 17(3), pp. 227–242 (2010).

96. Carro-Calvo, L., Salcedo-Sanz, S., Ortiz-Garc, G. and Portilla-Figueras, A. "An incremental-encoding evolutionary algorithm for color reduction in images", Integrated Computer-Aided Engineering, 17(3), pp. 261–269 (2010).
97. Baraldi, P., Canesi, R., Zio, E., Seraoui, R. and Chevalier, R. "Genetic algorithm-based wrapper approach for grouping condition monitoring signals of nuclear power plant components", Integrated Computer-Aided Engineering, 18(3), pp. 221–234 (2011).
98. Chabuk, T., Reggia, J.A., Lohn, J. and Linden, D. "Causally-guided evolutionary optimization and its application to antenna array design", Integrated Computer-Aided Engineering, 19(2), pp. 111–124 (2012).
99. Plevris, V. and Papadrakakis, M. "A hybrid particle swarm — gradient algorithm for global structural optimization", Computer-Aided Civil and Infrastructure Engineering, 26(1), pp. 48–68 (2011).
100. Pierce, S., Worden, K. and Manson, G. "A novel information-gap technique to assess reliability of neural network-based damage detection", Journal of Sound and Vibration, 293(1-2), pp. 96–111 (2006).
101. Franco, G., Betti, R. and Luş, H. "Identification of structural systems using an evolutionary strategy", Journal of Engineering Mechanics, 130(10), pp. 1125–1139 (2004).
102. Raich, A. and Liszkai, T. "Improving the performance of structural damage detection methods using advanced genetic algorithms", Journal of Structural Engineering, 133(6), pp. 449–461 (2007).
103. Zhao, J., Ivan, J. and DeWolf, J. "Structural damage detection using artificial neural networks", Journal of Infrastructure Systems, 4(3), pp. 93–101 (1998).
104. Adeli, H. and Park, H.S. "Counterpropagation neural network in structural engineering", Journal of Structural Engineering, 121(8), pp. 1205–1212 ASCE (1995).
105. Sirca, G. and Adeli, H. "Neural network model for uplift load capacity of metal roof panels", Journal of Structural Engineering, 127(11), pp. 1276–1285 (2001).

106. Bani-Hani, K., Ghaboussi, J. and Schneider, S. "Experimental study of identification and control of structures using neural network. part 1: identification", Earthquake Engineering and Structural Dynamics, 28(9), pp. 995–1018 (1999).
107. Hung, S.L. and Adeli, H. "Object-oriented back propagation and its application to structural design", Neurocomputing, 6(1), pp. 45–55 (1994).
108. Zang, C. and Imregun, M. "Structural damage detection using artificial neural networks and measured data reduced via principal component projection", Journal of Sound and Vibration, 242(5), pp. 813–827 (2001).
109. Hung, S. and Kao, C. "Structural damage detection using the optimal weights of the approximating artificial neural networks", Earthquake Engineering and Structural Dynamics, 31(2), pp. 217–234 (2002).
110. Ni, Y., Wang, B. and Ko, J. "Constructing input vectors to neural networks for structural damage identification", Smart Materials and Structures, 11(6), pp. 825–833 (2002).
111. Huang, C.S., Hung, S., Wen, C. and Tu, T. "A neural network approach for structural identification and diagnosis of a building from seismic response data", Earthquake Engineering and Structural Dynamics, 32(2), pp. 187–206 (2003).
112. Hao, H. and Xia, Y. "Vibration-based damage detection of structures by genetic algorithm", Journal of Computing in Civil Engineering, 16(3), pp. 222–229 (2002).
113. Perry, M., Koh, C. and Choo, Y. "Modified genetic algorithm strategy for structural identification", Computers & Structures, 84(8–9), pp. 529–540 (2006).
114. Yen, G. "Identification and control of large structures using neural networks", Computers & Structures, 52(5), pp. 859–870 (1994).
115. Adeli, H. and Karim, A. "Fuzzy-wavelet RBFNN model for freeway incident detection", Journal of Transportation Engineering, 126(6), pp. 464–471 (2000).
116. Karim, A. and Adeli, H. "Incident detection algorithm using wavelet energy representation of traffic patterns", Journal of Transportation Engineering, 128(3), pp. 232–242 (2002).

117. Karim, A. and Adeli, H. "Fast automatic incident detection on urban and rural freeways using wavelet energy algorithm", Journal of Transportation Engineering, 129(1), pp. 57–68 (2003).
118. Ghosh-Dastidar, S., Adeli, H. and Dadmehr, N. "Principal component analysis-enhanced cosine radial basis function neural network for robust epilepsy and seizure detection", IEEE Transactions on Biomedical Engineering, 55(2), p. 512 (2008).
119. Junfei, Q. and Honggui, H. "A repair algorithm for radial basis function neural network with application to chemical oxygen demand modeling", International Journal of Neural Systems, 20(1), pp. 63–74 (2010).
120. Patrinos, P., Alexandridis, A., Ninos, K. and Sarimveis, H. "Variable selection in nonlinear modeling based on RBF networks and evolutionary computation", International Journal of Neural Systems, 20(5), pp. 365–379 (2010).
121. Wu, D., Warwick, K., Ma, Z., Gasson, M.N., Burgess, J.G., Pan, S. and Aziz, T.Z. "Prediction of Parkinson's disease tremor onset using radial basis function neural network based on particle swarm optimization", International Journal of Neural Systems, 20(2), pp. 109–116 (2010).
122. Masri, S.F., Nakamura, M., Chassiakos, A.G. and Caughey, T.K. "Neural network approach to the detection of changes in structural parameters", Journal of Engineering Mechanics, 122(4), pp. 350–360 (1996).
123. Hung, S.L. and Adeli, H. "Parallel backpropagation learning algorithms on cray Y-MP8/864 supercomputer", Neurocomputing, 5(6), pp. 287–302 (1993).
124. Zapico, J., Worden, K. and Molina, F. "Vibration-based damage assessment in steel frames using neural networks", Smart Materials and Structures, 10(3), pp. 553–559 (2001).
125. Vaidyanathan, C., Kamatchi, P. and Ravichandran, R. "Artificial neural networks for predicting the response of structural systems with viscoelastic dampers", Computer-Aided Civil and Infrastructure Engineering, 20(4), pp. 280–291 (2005).
126. Na, C., Kim, S. and Kwak, H. "Structural damage evaluation using genetic algorithm", Journal of Sound and Vibration, 330(12), pp. 2772–2783 (2011).

127. Marano, G., Quaranta, G. and Monti, G. "Modified genetic algorithm for the dynamic identification of structural systems using incomplete measurements", Computer-Aided Civil and Infrastructure Engineering, 26(2), pp. 92–110 (2011).

128. Pandey, P. and Barai, S. "Multilayer perceptron in damage detection of bridge structures", Computers & Structures, 54(4), pp. 597–608 (1995).

129. Huang, C.C. and Loh, C. "Nonlinear identification of dynamic systems using neural networks", Computer-Aided Civil and Infrastructure Engineering, 16(1), pp. 28–41 (2001).

130. Wu, Z., Xu, B. and Yokoyama, K. "Decentralized parametric damage detection based on neural networks", Computer-Aided Civil and Infrastructure Engineering, 17(3), pp. 175–184 (2002).

131. Chen, C. "Structural identification from field measurement data using a neural network", Smart Materials and Structures, 14(3), pp. S104–S115 Institute of Physics (2005).

132. Xu, H. and Humar, J. "Damage detection in a girder bridge by artificial neural network technique", Computer-Aided Civil and Infrastructure Engineering, 21(6), pp. 450–464 (2006).

133. Jafarkhani, R. and Masri, S.F. "Finite element model updating using evolutionary strategy for damage detection", Computer-Aided Civil and Infrastructure Engineering, 26(3), pp. 207–224 (2011).

134. Mosquera, V., Smyth, A. and Betti, R. "Rapid evaluation and damage assessment of instrumented highway bridges", Earthquake Engineering and Structural Dynamics, 41(4), pp. 755–774 (2012).

135. Kang, F., Li, J. and Xu, Q. "Structural inverse analysis by hybrid simplex artificial bee colony algorithms", Computers & Structures, 87(13–14), pp. 861–870 (2009).

136. Karim, A. and Adeli, H. "Comparison of the fuzzy — wavelet RBFNN freeway incident detection model with the California algorithm", Journal of Transportation Engineering, 128(1), pp. 21–30 (2002).

137. Karim, A. and Adeli, H. "Radial basis function neural network for work zone capacity and queue estimation", Journal of Transportation Engineering, 129(5), pp. 494–503 (2003).

138. Adeli, H. and Ghosh-Dastidar, S. "Mesoscopic-wavelet freeway work zone flow and congestion feature extraction model", Journal of Transportation Engineering, 130(1), pp. 94–103 (2004).
139. Ghosh-Dastidar, S. and Adeli, H. "Neural network-wavelet microsimulation model for delay and queue length estimation at freeway work zones", Journal of Transportation Engineering, 132(4), pp. 331–341 (2006).
140. Boto-Giralda, D., Díaz-Pernas, F.J, González-Ortega, D., Díez-Higuera, J.F., Antón-Rodríguez, M. and Martínez-Zarzuela, M. "Wavelet-based denoising for traffic volume time series forecasting with self-organizing neural networks", Computer-Aided Civil and Infrastructure Engineering, 259(7), pp. 530–545 (2010).
141. Ghosh, B., Basu, B. and O'Mahony, M. "Random process model for traffic flow using a wavelet-Bayesian hierarchical technique", Computer-Aided Civil and Infrastructure Engineering, 25(8), pp. 613–624 (2010).
142. Zhou, Z. and Adeli, H. "Time-frequency signal analysis of earthquake records using Mexican hat wavelets", Computer-Aided Civil and Infrastructure Engineering, 18(5), pp. 379–389 (2003).
143. Adeli, H. and Kim, H. "Wavelet-hybrid feedback-least mean square algorithm for robust control of structures", Journal of Structural Engineering, 130(1), pp. 128–137 ASCE (2004).
144. Jiang, X. and Adeli, H. "Wavelet packet-autocorrelation function method for traffic flow pattern analysis", Computer-Aided Civil and Infrastructure Engineering, 19(5), pp. 324–337 (2004).
145. Kim, H. and Adeli, H. "Hybrid control of smart structures using a novel wavelet-based algorithm", Computer-Aided Civil and Infrastructure Engineering, 20(1), pp. 7–22 (2005).
146. Jiang, X. and Adeli, H. "Dynamic fuzzy wavelet neuroemulator for nonlinear control of irregular highrise building structures", International Journal for Numerical Methods in Engineering, 74(7), pp. 1045–1066 (2008).
147. Liew, K. and Wang, Q. "Application of wavelet theory for crack identification in structures", Journal of Engineering Mechanics, 124(2), pp. 152–157 (1998).

148. Pan, T. and Lee, C. "Application of wavelet theory to identify yielding in seismic response of bi-linear structures", Earthquake Engineering and Structural Dynamics, 31(2), pp. 379–398 (2002).

149. Bao, C., Hao, H., Li, Z. and Zhu, X. "Time-varying system identification using a newly improved HHT algorithm", Computers & Structures, 87(23–24), pp. 1611–1623 (2009).

150. Kijewski, T. and Kareem, A. "Wavelet transforms for system identification: considerations for civil engineering applications", Computer-Aided Civil and Infrastructure Engineering, 18(5), pp. 339–355 (2003).

151. Qiao, L., Esmaeily, A. and Melhem, H.G. "Signal pattern-recognition for damage diagnosis in structures", Computer-Aided Civil and Infrastructure Engineering, 27(9), (2012), http://dx.doi.org/10.1111/j.1467- 8667.2012.00766.x.

152. Huang, C.S., Hung, S., Lin, C. and Su, W. "A wavelet-based approach to identifying structural modal parameters from seismic response and free vibration data", Computer-Aided Civil and Infrastructure Engineering, 20(6), pp. 408–423 (2005).

153. Tee, K., Koh, C. and Quek, S. "Substructural first- and second-order model identification for structural damage assessment", Earthquake Engineering and Structural Dynamics, 34(15), pp. 1755–1775 (2005).

154. Banerjee, B., Roy, D. and Vasu, R.M. "Self-regularized pseudo timemarching schemes for structural system identification with static measurements", International Journal for Numerical Methods in Engineering, 82(7), pp. 896–916 (2010).

155. Sepe, V., Capecchi, D. and De Angelis, M. "Modal model identification of structures under unmeasured seismic excitations", Earthquake Engineering and Structural Dynamics, 34(7), pp. 807–824 (2005).

156. Yuen, K. and Katafygiotis, L. "Model updating using noisy response measurements without knowledge of the input spectrum", Earthquake Engineering and Structural Dynamics, 34(2), pp. 167–187 (2005).

157. Loh, C., Mao, C., Huang, J. and Pan, T. "System identification and damage evaluation of degrading hysteresis of reinforced concrete frames", Earthquake Engineering and Structural Dynamics, 40(6), pp. 623–640 (2010).

158. Hazra, B., Sadhu, A., Roffel, J. and Narasimhan, "Hybrid time-frequency blind source separation towards ambient system identification of structures", Computer-Aided Civil and Infrastructure Engineering, 27(5), pp. 314–332 (2012).

159. Hwang, J., Kareem, A. and Kim, W. "Estimation of modal loads using structural response", Journal of Sound and Vibration, 326(3–5), pp. 522–539 (2009).

160. Kijewski-Correa, T., Kwon, D., Kareem, A., Bentz, A., Guo, Y., Bobby, S. and Abdelrazaq, A. "SmartSync: an integrated real-time structural health monitoring and structural identification system for tall buildings", Journal of Structural Engineering, (2012), http://dx.doi.org/10.1061/(ASCE)ST.1943-541X.0000560.

161. Lee, Z., Wu, T. and Loh, C. "System identification on the seismic behavior of an isolated bridge", Earthquake Engineering and Structural Dynamics, 32(12), pp. 1797–1812 (2003).

162. Yan, B. and Miyamoto, A. "A comparative study of modal parameter identification based on wavelet and Hilbert-Huang transforms", ComputerAideyan and Civil and Infrastructure Engineering, 21(1), pp. 9–23 (2006).

163. Douka, E., Loutridis, S. and Trochidis, A. "Crack identification in plates using wavelet analysis", Journal of Sound and Vibration, 270(1–2), pp. 279–295 (2004).

164. Kim, B., Kim, H. and Park, T. "Nondestructive damage evaluation of plates using multi-resolution analysis of two-dimensional Haar wavelet", Journal of Sound and Vibration, 292(1–2), pp. 82–104 (2006).

165. Rucka, M. and Wilde, K. "Application of continuous wavelet transform in vibration based damage detection method for beams and plates", Journal of Sound and Vibration, 297(3–5), pp. 536–550 (2006).

166. Castro, E., García-Hernandez, M. and Gallego, A. "Damage detection in rods by means of the wavelet analysis of vibrations: influence of the mode order", Journal of Sound and Vibration, 296(4–5), pp. 1028–1038 (2006).

167. Le, T. and Argoul, P. "Continuous wavelet transform for modal identification using free decay response", Journal of Sound and Vibration, 277(1–2), pp. 73–100 (2004).

168. Chakraborty, A., Basu, B. and Mitra, M. "Identification of modal parameters of a MDOF system by modified L-P wavelet packets", Journal of Sound and Vibration, 295(3–5), pp. 827–837 (2006).

169. Chen, S., Liu, J. and Lai, H. "Wavelet analysis for identification of damping ratios and natural frequencies", Journal of Sound and Vibration, 323(1–2), pp. 130–147 (2009).

170. Ching, J. and Glaser, S. "Tracking rapidly changing dynamical systems using a non-parametric statistical method based on wavelets", Earthquake Engineering and Structural Dynamics, 32(15), pp. 2377–2406 (2003).

171. He, Z., You, X., Zhou, L., Cheung, Y. and Tang, Y.Y. "Writer identification using fractal dimension of wavelet subbands in Gabor domain", Integrated Computer-Aided Engineering, 17(2), pp. 157–165 (2010).

172. Schoefs, F., Yáñez-Godoy, H. and Lanata, F. "Polynomial chaos representation for identification of mechanical characteristics of instrumented structures", Computer-Aided Civil and Infrastructure Engineering, 26(3), pp. 173–189 (2011).

173. Li, H., Huang, Y., Ou, J. and Bao, Y. "Fractal dimension-based damage detection method for beams with a uniform cross-section", ComputerAided Civil and Infrastructure Engineering, 26(3), pp. 190–206 (2011).

174. Adeli, H. and Hung, S.L., Machine Learning-Neural Networks, Genetic Algorithms, and Fuzzy Systems, John Wiley and Sons, New York (1995).

175. Jiang, X. and Adeli, H. "Fuzzy clustering approach for accurate embedding dimension identification in chaotic time series", Integrated ComputerAided Engineering, 10(3), pp. 287–302 (2003).

176. Ma, Z.M., Zhang, F., Yan, L. and Cheng, J. "Extracting knowledge from fuzzy relational databases with description logic", Integrated Computer-Aided Engineering, 18(2), pp. 181–200 (2011).

177. Panakkat, A. and Adeli, H. "Neural network models for earthquake magnitude prediction using multiple seismicity indicators", International Journal of Neural Systems, 17(1), pp. 13–33 (2007).

178. Reuter, U. "A fuzzy approach for modeling non-stochastic heterogeneous data in engineering based on cluster analysis",

Integrated Computer-Aided Engineering, 18(3), pp. 281–289 (2011).

179. Sarma, K.C. and Adeli, H. "Bi-level parallel genetic algorithms for optimization of large steel structures", Computer-Aided Civil and Infrastructure Engineering, 16(5), pp. 295–304 (2001).

180. Gonzalez-Olvera, M.A., Gallardo-Hernandez, A.G., Tang, Y., RevillaMonsalve, M.C. and Islas-Andrade, S. "A discrete-time recurrent neurofuzzy network for black-box modeling of insulin dynamics in diabetic Type-1 patients", International Journal of Neural Systems, 20(2), pp. 149–158 (2010).

181. Wang, N., Joo, M., Meng, X.Y. and Li, X. "An online self-organizing scheme for parsimonious and accurate fuzzy neural networks", International Journal of Neural Systems, 20(5), pp. 389–403 (2010).

182. Koprinkova-Hriatova, P. "Backpropagation through time training of a neuro-fuzzy controller", International Journal of Neural Systems, 20(5), pp. 421–428 (2010).

183. Yu, W. and Li, X. "Automated nonlinear system modeling with multiple fuzzy neural networks and kernel smoothing", International Journal of Neural Systems, 20(5), pp. 429–435 (2010).

184. Theodoridis, D.C., Boutalis, Y.S. and Christodoulou, M.A. "indirect adaptive control of unknown multi variable nonlinear systems with parametric and dynamic uncertainties using a new neuro-fuzzy system description", International Journal of Neural Systems, 20(2), pp. 129–148 (2010).

185. Bianchini, A. and Bandini, P. "Prediction of pavement performance through neuro-fuzzy reasoning", Computer-Aided Civil and Infrastructure Engineering, 25(1), pp. 39–54 (2010).

186. Iglesias, J.A., Angelov, P., Ledezma, A. and Sanchis, A. "Human activity recognition based on evolving fuzzy systems", International Journal of Neural Systems, 20(5), pp. 355–364 (2010).

187. Jiang, X. and Adeli, H. "Neuro-genetic algorithm for nonlinear active control of highrise buildings", International Journal for Numerical Methods in Engineering, 75(7), pp. 770–786 (2008).

188. Wadia-Fascetti, S. and Gunes, B. "Earthquake response spectra models incorporating fuzzy logic with statistics", Computer-Aided Civil and Infrastructure Engineering, 15(2), pp. 134–146 (2000).

189. Adeli, H. and Jiang, X. "Dynamic fuzzy wavelet neural network model for structural system identification", Journal of Structural Engineering, 132(1), pp. 102–111 (2006).

190. Jiang, X. and Adeli, H. "Dynamic wavelet neural network for nonlinear identification of highrise buildings", Computer-Aided Civil and Infrastructure Engineering, 20(5), pp. 316–330 (2005).

191. Osornio-Rios, R.A., Amezquita-Sanchez, J.P., Romero-Troncoso, R.J. and Garcia-Perez, A. "MUSIC-neural network analysis for locating structural damage in truss-type structures by means of vibrations", Computer-Aided Civil and Infrastructure Engineering, 27(9), pp. 687–698 (2012).

Chapter 9

Fast Dynamic Control of Damaged Historical Buildings: A New Useful Approach for Structural Health Monitoring after an Earthquake

Sergio Vincenzo Calcina, Luca Piroddi, and Gaetano Ranieri

Department of Civil Engineering, University of Cagliari, Environmental Engineering and Architecture (DICAAR), Via Marengo 2, 09123 Cagliari, Italy.

ABSTRACT

The structures damage conditions assessment requires numerous precautions to ensure the safety of people during site visits and inspections. Among several methods providing useful information about the conservation status of the structures, dynamic monitoring techniques are suitable to retrieve the global behavior of the buildings.

The anomalous features diagnosis of the structural dynamic response is an index of alterations of the material state and, in the worst cases, is related to the presence of damaged structural elements. This paper proposes the use of remote control systems for the structural evaluation of the damage state of buildings and describes the results achieved in an interesting application: the experimental dynamic analysis carried out on the inaccessible damaged bell tower of the Church of Santi Giacomo and Filippo in Mirandola (Italy). The study is based on observations performed using the IBIS-S ground-based radar interferometer to remotely measure the displacements of several elements of the building above 0.01 mm amplitude. This totally noninvasive and nondestructive approach has proved to be reliably implemented as a useful method to structural health monitoring procedures and especially for extensive and fast inspection analyses aiming at the first evaluation of the damage level and the soundness of slender buildings after earthquakes.

INTRODUCTION

Structural damage identification is a fundamental element following an earthquake. A correct definition of the damage state of buildings allows us to establish technical procedures and operational standards for safeguarding the structures, aimed at restoring their original conditions.

Structural Health Monitoring techniques (SHM) make it possible to deduce the presence of lesions and estimate the severity of the damage to the structures by measuring the mode of vibration of the buildings. This is possible because the dynamic response of structures is strongly influenced by the conservation state of materials and by structural lesions. The scientific literature includes many methods of structural dynamic response evaluation and algorithms for the extraction of main modal parameters [1, 2]. Several experimental configurations are classified in function of the number of measured output signals (number of sensors) and of the type of sources used to energize the structures. In this context, there are several experimental layouts to modal parameters identification. Specifically, it is possible to differentiate between forced vibration tests (i.e., using mechanical shaker or vibrodines), free vibration tests (i.e., by means of falling objects inside or outside the buildings-impulsive signals), and ambient vibration tests (i.e., using random sources as wind and artificial noise) [3]. The last methods are very useful for assessing dynamic behavior and for a rapid evaluation of the true conditions of existing damaged structures after an earthquake. However, the most common data acquisition techniques use in-contact sensors. Velocimeters or accelerometers arrays are placed inside the building and are used to record time series of ambient vibration. In the last decade, advances in the field of Ground Based Remote Sensing assure remote data acquisition and real-time monitoring of vibrations in critical conditions, as well as dynamic control of severely damaged structures after earthquakes. In this paper, the authors propose a procedure for assessing the state of damage to structures in areas affected by earthquakes. This approach is founded on remote monitoring techniques of mechanical vibrations by means of interferometric surveys. This method allows us to ensure maximum safety conditions during the monitoring of potentially damaged structures. In fact, the assessment of the structural damages during the next phase is extremely delicate due to precarious stability conditions of structures and due to possible further aftershocks during technical inspections by specialists. For this reason, a new application of this proposed operational approach will be discussed in the case of the Emilia earthquake (Italy) for the stability control of the bell tower of San Giacomo Roncole using IBIS-S ground-based microwave interferometer.

Instrumental Features of the IBIS-S Technology

Microwave interferometry has recently emerged as a new technology, specifically suitable to remotely measure the vibration response of structures. Several authors observe that the interferometric technique has proven to be a useful remote sensing tool for vibration measurements of structures, such as architectural heritage structures [4–6], engineering infrastructures [7, 8], and vibrating stay cables [9, 10]. This radar technology is founded on the combined use of high-resolution waveforms [11] and interferometric technique [12]. This technique is implemented to compute the displacement of each target through the phase shift Δ of the back-scattered microwave signals collected at different time intervals. The displacement along the radar line of sight d_r is computed from the phase-shift with the following simple equation (1):

$$dr = -\frac{\lambda}{4\pi}\Delta\theta, \tag{1}$$

Where λ is the wavelength of the signal [13]. The main advantages of the interferometric radar for structural health monitoring of civil infrastructures are high accuracy of measurements (sensitivity of 0.01–0.02 mm) and high spatial resolution, simultaneous monitoring of several targets within the sensor applicable distance, independence from daylight and weather conditions for short recordings, portability, and quick set-up time. The sensor module is provided with two-horn antennas (−3 dB beam width of 17° and 15° in the horizontal and vertical plane) and transmits the electromagnetic signals in Ku frequency band from 17.1 to 17.3 GHz, receiving the echoes from several targets placed inside the radar scenario.

The acquired time history s (t) is the displacement of the target projected along the sensor-target line of sight (LOS). The LOS displacement s (t) is linked to the displacements components V (t) and H (t), in vertical and in horizontal plane respectively, through the following equation (2):

$$s(t) = V(t)\sin(\beta) + H(t)\cos(\beta), \tag{2}$$

With β being the LOS angle computed from the horizontal axis. The interferometric survey may thus be considered a powerful tool to evaluate the deformation related to dynamic applied loads and to evaluate the stiffness of structures.

Description and Method

The test site was chosen in order to analyze a strongly damaged but not totally crushed bell tower, for which remote sensing approach was the only compatible one with safety conditions. The experimental settings were designed with the task to have comparable results from two main acquisition distances from the target and two different direction projections of measured displacement.

The Tower: A Brief Historical Overview and Description of the Damage

The Emilia earthquake has caused numerous damages to the structures. The buildings most affected by the earthquake are industrial structures (factories and storage magazines) and cultural heritage buildings (churches, bell towers, civic towers, etc.), whose geometric and constructive features have contributed to increasing their vulnerability in case of earthquakes. In this work, we present the study performed for the bell tower of San Giacomo Roncole. The actual structure was built from 1771 to 1774 with the same materials of the previous Carmelite Convent of Santa Maria delle Grazie della Galeazza [14]. The building is located few metres away from the singular nave of the church. The structure replaced an earlier tower, adjacent to the body of the church, and demolished by the damage caused by the vibrations of the bells. The new tower is a masonry structure located at southwest of the church. The tower reaches a height of about 36 meters with a square base (5 meters per side). After the earthquakes of the 20th and 29th May, 2012, (magnitude m_1=5.9 and ,m_1=5.8 resp.) the building sustained serious damage to the base, highlighted by displacements of about 2 cm toward the southwest side and of 1 cm on the southeast side. In other words, after the earthquake, the structure was shifted at the base, assuming an eccentric position. Also there are various fractures on the first section of the tower horizontally and diagonally crossing its perimeter walls, where a rigid motion on the order of centimeters was measured. The top of the structure collapsed at about 9 meters from the base toward the southwest side (see Figure 1).

Figure 1: Some images of the damaged bell tower. (a) Southwest side of the building. (b) Panoramic view. (c) Cracks at the base of the tower. (d) Collapsed top.

Sensor Installation and Processing of Data

The tower of San Giacomo Roncole is a slim structure with a shape characterized by several natural corners (see Figure 2). Consequently, the measurements are performed without artificial corner reflectors installed on the body surface of the structure. This is an essential condition because the structure is seriously damaged. Dynamic surveys are carried out using three different configurations. In the first one, the IBIS-S system was placed in front of the tower in the southeast direction in order to monitor the dynamic behavior of the full body of the structure with 30° LOS angle, computed for the emitting cone axis of the radar sensor. The second radar station was placed in the southwest direction (30° LOS angle). The last acquisition was performed with the IBIS-S system located under the tower (70° LOS angle). The acquisition geometry was selected in order to reduce the effects of all possible noise sources, such as suspended cables and other metal structures or trees and vegetation inside the radar profile. Each acquisition had a duration of about 10 min with a sampling frequency of 200 Hz. The vibrations of the building were monitored in real time during field operations. The Fourier spectra of signals were provided by processing the acquired data with IBIS Data viewer and Matlab software in order to assess the main frequencies of the structure. Before of the spectral analysis, the displacement data were detrended and tapered with Hamming function to reduce leakage effects. We have applied the Frequency Domain Decomposition technique [15] based on singular values decomposition of the power spectral density matrix to identify the structural dynamic response of the tower. By using this decomposition method, main modal parameters can be identified with high accuracy even in the case of strong noise contamination of the signals.

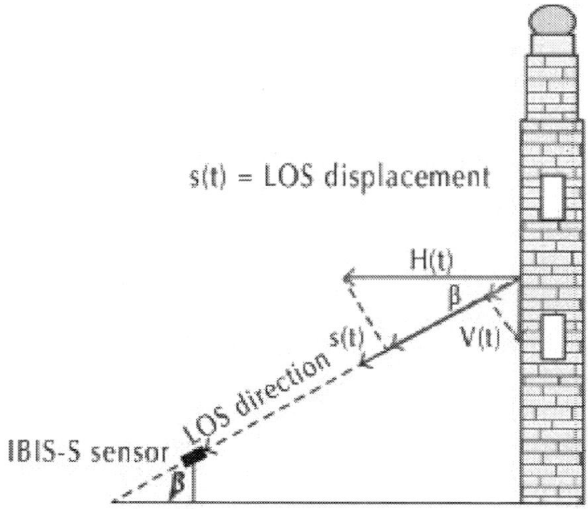

Figure 2: (a) Installation of the IBIS-S system in several acquisition stations placed at 25 m (St 01) and 23 m (St 02) from the base of the tower with LOS angle of 30° (central beam) and at 8 m (St 03) with LOS angle of 70° (central beam). (b) Radar line of sight displacement and true displacement vector. (c) Image of the IBIS-S system used during the experiments.

RESULTS AND DISCUSSION

The range bins with high signal-to-noise ratio (SNR) along the radar power profile are associated with different natural back-scatterers located inside the radar scenario. However, the signal-to-noise ratio of the radar response from this structure is high and the structure is isolated inside the radar scenario. Thus several range bins may be selected to be analyzed. Five range bins of the radar power profile are selected to estimate fundamental vibration properties of the structure by means of the FDD decomposition technique. The radar bins are selected in function of the distance along the radar line of sight between each radar-bin and the IBIS-S sensor. These range bins correspond to five reflective points placed on the bell tower facade. This operation is easily performed knowing the geometric characteristics of the tower (the height of several architectonic elements such as windows and other ledges), the horizontal distance of the radar station, and the inclination of the IBIS-S head. The displacements retrieved from radar data are shown in Figure 3. These time series have been obtained observing the tower from a position to the base of the tower, at a distance of 23 metres (configuration 2). During all measurements, a regular wind with moderate speed of 5.5 m/s blew.

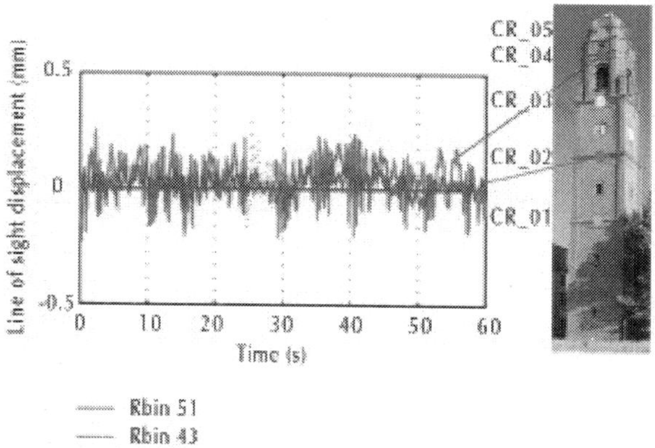

Figure 3: One-minute record acquired with the second configuration: LOS displacement retrieved from radar data as a function of the time. The time histories refer to two range bins selected from the radar power profile and correspond to different heights of the tower.

The calculated displacement spectra, obtained over the whole recorded ten minutes signal, are shown in Figure4. A band-pass filter (0.3 Hz–10 Hz) was applied to reduce the presence of disturbances linked to atmospheric effects and to vibration of the sensor itself. The influence of the first effect is prevalent on the measured phases and, as a consequence, on the estimated displacements with long period (0.1–0.2 Hz). Another effect is caused by the vibration of the measurement tool. Luzi et al. [16] observe that the measured resonant frequency of the system ranges within 10 Hz ± 2 Hz for typical working conditions. Then a low-pass filtering may be applied in order to assure that the measured vibrations are not imputable to the natural response of the tripod-head sensor system. All amplitude spectra are characterized by a clear harmonic component at 0.96 Hz (corresponding period of 1.04 s) whose amplitude increases with the height of the analyzed range bin. This behavior agrees with the expected amplitude variation associated at the first natural mode (simple bending). This periodicity cannot be directly retrieved from all displacement time series but only for the 50th range bin corresponding to the reflection from the height of 30 meters on the tower (see Figure 5).

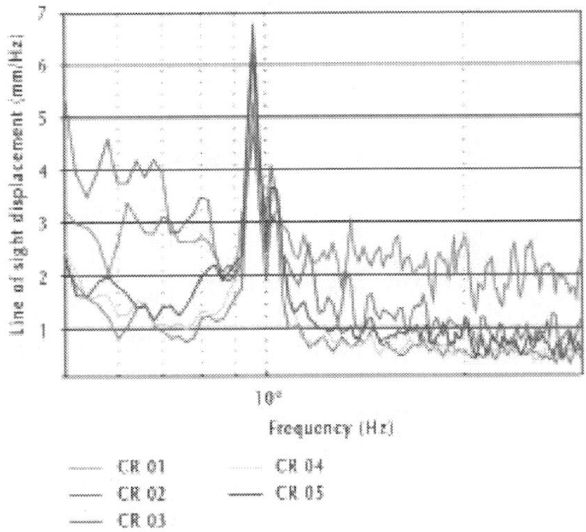

Figure 4: Displacement spectra calculated by using the entire 10-minute record, for five different radar-bins with variable SNR achieved from data acquisition with configuration 1.

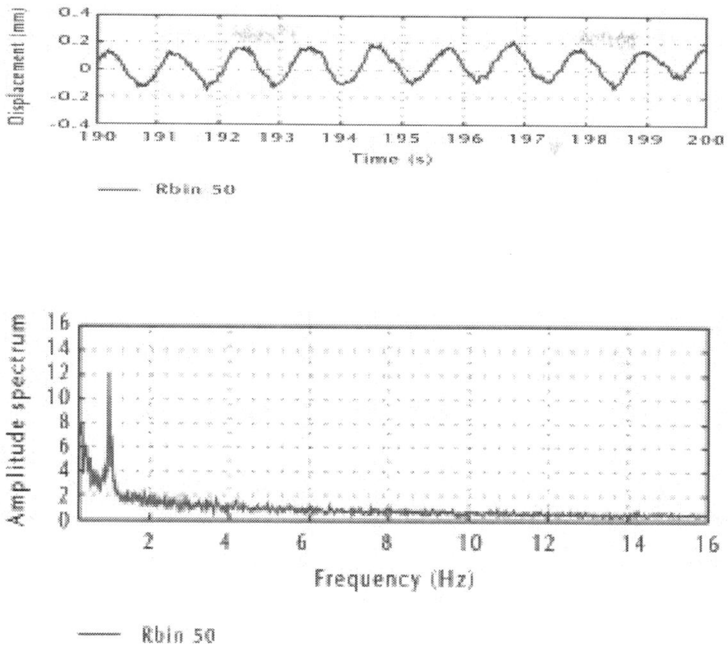

Figure 5: (a) Sinusoidal trend of the displacement record obtained for the 50th range bin (30 metres height). (b) Frequency analysis of the same time-history.

The obtained value of 0.96 Hz could be influenced by the presence of structural damage caused by the earthquake. This value could be compared with the theoretical one achieved through numerical modeling of the building before the damage. However, even without a direct model, several empirical relationships available for masonry structures, like those proposed by the Italian Building Code NTC-2008, the Spanish National Code NSCE-02, and by literature studies [17], are used to assess the fundamental frequency of vibration. The obtained results range between 1.4 and 1.5 Hz. The significant difference (31%–36%) between the experimental value and the estimated values could be imputed to the damage caused by the earthquake. It is generally recognized that the period of vibration grows while increasing the mass of the vibrating system and while reducing the stiffness.

The radar measurements allow the evaluation of the dynamic behavior (amplitude of vibration) of several parts of the building located at different heights. In this case we have monitored the structure from different perspectives and we have compared the experimental mode shapes obtained for the IBIS-S configurations 1 and 2. The mode shape retrieved by the configuration 3 was not used for this analysis because the displacements measured in LOS direction are not comparable with the others. In fact, in this configuration, the microwave sensor is more inclined (70°) and is placed at short distance from the tower (8 metres). Measured displacements along two directions are projected along the horizontal plane. The experimental mode shapes show a similar trend, as we can see in Figure 6. We observe that maximum values are measured from St 02 IBIS-S station and are characterized by amplitude ranges from 0.05 to 0.2 mm. The fundamental mode of vibration is a bending mode with the same frequency of 0.96 Hz in both directions of measurement. Anomalous behavior is detected at the point placed at about 17 metres height with experimental measured displacements greater than those expected. At this level of the structure, we observe a significant increasing in the amplitude of the measured displacements with respect to the lower level (located at 11 metres height above the ground surface). There is an increase of 100% (acquisition station 1), moving from 0.04 mm to 0.08 mm of maximum amplitude at the first structural vibration frequency and an increase of 80% (acquisition station 2), from 0.05 mm to 0.09 mm, which are not attributable to geometric differences in the two building levels or to largely variable construction features of the elevating structure. Therefore, this trend could be reliably correlated with the high structural damages of the building at the middle level. Moreover, the larger displacements obtained in the second acquisition are justified by the major shift recorded in this direction (about 2 cm) due to seismic action.

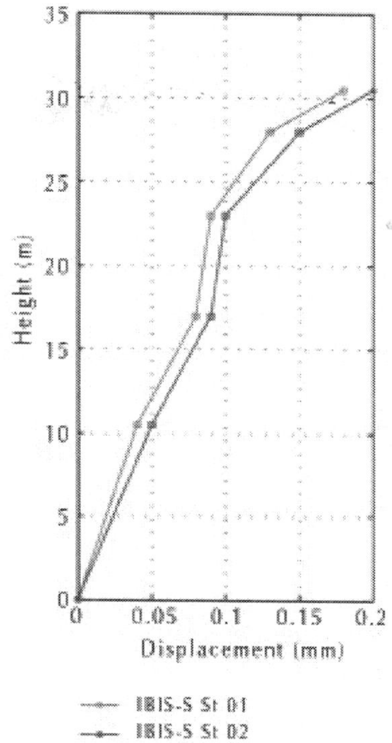

Figure 6: Comparison between the experimental mode shapes retrieved from two points of view of the IBIS-S system.

CONCLUSIONS

This paper proposes the use of the ground-based microwave interferometry for remote sensing of vibrations of structures aimed at providing useful information for specialists about the damage state of structures in areas affected by earthquakes. For this reason, this work is the first application of ground-based radar interferometry to the study of an inaccessible damaged building. The authors report the results of the experimental dynamic test aimed at estimating the vibrations of an historical bell tower in an urban environment, performed through ambient vibration testing. Also, the basic principle of the interferometric technique has been briefly resumed and the main experimental data obtained for the vibration features of the structure have been critically discussed to identify anomalous dynamic behaviors linked to structural damages. In general, the proposed method can estimate displacements ranging from a few microns up to several millimeters for large structures, ranging from thin and tall structures to other types of buildings (towers, skyscrapers, and bridges). The possibility of working remotely makes this approach suitable for the dynamic control of buildings that have reported structural damage after an earthquake, especially for civil structures of strategic interest during the emergency and for cultural heritage buildings which represent the history and the social identity of a community.

CONFLICT OF INTERESTS

The authors declare that there is no conflict of interests regarding the publication of this paper titled "Fast dynamic control of damaged historical buildings: a new useful approach for Structural Health Monitoring after an earthquake."

ACKNOWLEDGMENTS

Sergio Vincenzo Calcina gratefully acknowledges Sardinia Regional Government for the financial support of her Ph.D. scholarship (P.O.R. Sardegna F.S.E. Operational Programme of the Autonomous Region of Sardinia, European Social Fund 2007–2013—Axis IV Human Resources, Objective l.3, Line of Activity l.3.1.). Luca Piroddi gratefully acknowledges Sardinia Regional Government for the financial support (P.O.R. Sardegna F.S.E. Operational Programme of the Autonomous Region of Sardinia, European Social Fund 2007–2013—Axis IV Human Resources, Objective l.3, Line of Activity l.3.1 "Avviso di chiamata per il finanziamento di Assegni di Ricerca").

REFERENCES

1. S. S. Ivanovic, M. D. Trifunac, and M. I. Todorovska, "Ambient vibration tests of structures: a review,"Bulletin of Indian Society of Earthquake Technology, vol. 37, no. 4, article 407, pp. 165–197, 2000.

2. H. Sohn, C. R. Farrar, F. M. Hemez et al., "A Review of structural health monitoring literature: 1996–2001," Los Alamos National Laboratory Report LA-13976-MS, Los Alamos, NM, USA, 2004.
3. N. M. M. Maia, J. M. M. Silva, J. He et al., Theoretical and Experimental Modal Analysis, Research Studies Press, New York, NY, USA, 1997.
4. M. Pieraccini, M. Fratini, F. Parrini, G. Pinelli, and C. Atzeni, "Dynamic survey of architectural heritage by high-speed microwave interferometry," IEEE Geoscience and Remote Sensing Letters, vol. 2, no. 1, pp. 28–30, 2005.
5. M. Pieraccini, M. Fratini, D. Dei, and C. Atzeni, "Structural testing of historical heritage site towers by microwave remote sensing," Journal of Cultural Heritage, vol. 10, no. 2, pp. 174–182, 2009.
6. C. Atzeni, A. Bicci, D. Dei, M. Fratini, and M. Pieraccini, "Remote survey of the leaning tower of pisa by interferometric sensing," IEEE Geoscience and Remote Sensing Letters, vol. 7, no. 1, pp. 185–189, 2010.
7. C. Gentile and G. Bernardini, "Output-only modal identification of a reinforced concrete bridge from radar-based measurements," NDT & E International, vol. 41, no. 7, pp. 544–553, 2008.
8. C. Gentile and G. Bernardini, "An interferometric radar for non-contact measurement of deflections on civil engineering structures: Laboratory and full-scale tests," Structure and Infrastructure Engineering, vol. 6, no. 5, pp. 521–534, 2010.
9. C. Gentile, "Application of microwave remote sensing to dynamic testing of stay-cables," Remote Sensing, vol. 2, no. 1, pp. 36–51, 2010.
10. C. Gentile, "Deflection measurement on vibrating stay cables by non-contact microwave interferometer," NDT & E International, vol. 43, no. 3, pp. 231–240, 2010.
11. D. R. Wehner, High-Resolution Radar, Artech House, London, UK, 2nd edition, 1995.
12. F. M. Henderson and A. J. Lewis, Manual of Remote Sensing. Principles and Applications of Imaging Radar, Wiley, New York, NY, USA, 1998.
13. J. D. Taylor, Ultra-Wideband Radar Technology, CRC Press, Boca Raton, Fla, USA, 2001.

14. E. Andreoli, San Giacomo Roncole: Raccolta Di Notizie Storiografiche Popolari, Mirandola Edizioni, Mirandola, Italy, 1987.
15. R. Brincker, L. Zhang, and P. Andersen, "Modal identification of output-only systems using frequency domain decomposition," Smart Materials and Structures, vol. 10, no. 3, pp. 441–445, 2001.
16. G. Luzi, O. Monserrat, and M. Crosetto, "Real aperture radar interferometry as a tool for buildings vibration monitoring: limits and potentials from an experimental study," in Proceedings of the 10th International Conference on Vibration Measurements by Laser and Non-contact Techniques (AIVELA '12), pp. 309–317, Ancona, Italy, 2012.
17. C. Rainieri and G. Fabbrocino, "Estimating the elastic period of Masonry towers," in Proceedings of the SEM IMAC 30th Conference, Society for Experimental Mechanics, Jacksonville, Fla, USA, 2012.

Chapter 10

A Review on Strengthening Steel Beams Using FRP under Fatigue

Mohamed Kamruzzaman, Mohd Zamin Jumaat, N. H. Ramli Sulong, and A. B. M. Saiful Islam

Department of Civil Engineering, University of Malaya, 50603 Kuala Lumpur, Malaysia

ABSTRACT

In recent decades, the application of fibre-reinforced polymer (FRP) composites for strengthening structural elements has become an efficient option to meet the increased cyclic loads or repair due to corrosion or fatigue cracking. Hence, the objective of this study is to explore the existing FRP reinforcing techniques to care for fatigue damaged structural steel elements. This study covers the surface treatment techniques, adhesive curing, and support conditions under cyclic loading including fatigue performance, crack propagation, and failure modes with finite element (FE) simulation of the steel bridge girders and structural elements. FRP strengthening composites delay initial cracking, reduce the crack growth rate, extend the fatigue life, and decrease the stiffness decay with residual deflection. Prestressed carbon fibre-reinforced polymer (CFRP) is the best strengthening option. End anchorage prevents debonding of the CRRP strips at the beam ends by reducing the local interfacial shear and peel stresses. Hybrid-joint, nanoadhesive, and carbon-flex can also be attractive for strengthening systems.

INTRODUCTION

The collapse of structural elements due to fatigue is extremely expensive and may be catastrophic in terms of human life and damage. The fatigue process can be defined as the accumulation of damage at a localized region as a result of cyclic loads, which leads to the formation of a crack that eventually propagates through the material. When a crack grows to a size for which the net-section is inadequate to carry the imposed load, rapid fracture takes place. Fatigue failure of steel girders is one of the most significant problems affecting the remaining service life of existing steel structures like bridges. The application of fibre-reinforced polymer (FRP) composites to increase the fatigue strength of damaged steel girders is a promising technique that offers an attractive substitute to traditional approaches like steel plating.

In recent decades, the strengthening of the steel components of structures has become essential for structural retrofits. FRP strengthening seems to be an effective alternative for such retrofitting. A survey of European railway administrations covering nearly 220,000 bridges across Europe [1] has shown that almost 22% of the bridges are constructed with metal. Among them, 28% of the metallic bridges are over 100 years old, and 40% are between 50 and 100 years old. The federal highway administration of the US Department of Transportation, 2005 [2], estimated that among the almost 200,000 metallic highway bridges in the US, around 40% are either "structurally deficient" or "functionally obsolete." Deterioration is mostly due to accidental vehicular impacts, cyclic loads. Bridge structures that support moving trains are subjected to high stresses caused by extreme vibrations and dynamic deflections that are far greater than ever before [2]. Moreover, the fatigue life of steel structures could be extensively reduced due to earthquake loadings [3]. The dynamic reaction of railway bridges is also influenced by various factors including the train speed, natural frequency of the bridge, and the bridge and carriage lengths [4]. The I-35W Bridge over the Mississippi river collapsed on 1 August 2007, resulting in 13 dead and 145 injuries, and damaged 111 vehicles. The National Transportation Safety Board [5] cited on the collapse of the I-35W Bridge that a design fault, improper maintenance, and additional weight, because of increasing traffic volume and heavy traffic, caused the fatigue failure. The fatigue damage reported in Japan has been caused mainly by cyclic secondary stress or distortion-induced stress [6].

Fibre-reinforced polymer possesses outstanding advantages as a structural material, including high strength, anticorrosion properties, and high durability and is able to restore the lost capacity of damaged structures [7, 8]. FRP sheets/strips are also effective in the strengthening of steel structural elements to extend their fatigue lifetime and reduce crack propagation [9–11] if galvanic corrosion is prevented and sufficient bond is provided [12, 13]. Recent strengthening projects in the United States, Switzerland, the United Kingdom, and Japan showed that there was a great potential for using carbon fibre-reinforced polymer (CFRP) to retrofit steel structural elements [14–16]. CFRP materials have been used for marine structures [17, 18] and aerospace parts [19] and have expanded to wood [20], masonry [21], steel/concrete composite structures [22], and so on. It is more frequently used for rehabilitation and strengthening of steel structures than other FRP materials due to its high strength. In any case, CFRP is very tolerant to fatigue damage [23, 24]. Basalt-fibre-reinforced polymers (BFRP) have been increasingly considered in civil infrastructures, because of low cost and their excellent chemical and mechanical properties [25, 26].

The Acton Bridge in England [27] and the Ashland Bridge and 7838.5S092 Bridge in the USA were retrofitted by applying CFRP elements to the bottom of the girders; the stress reduction on the original materials was observed to increase the fatigue life [28]. Several case studies of existing steel railway bridges are presented in [29, 30]. Monitoring data was used for dynamic response analysis and fatigue life evaluation. It was observed that the stringers and the cross beams run a high risk of fatigue damage. The weakest point in the reinforcing system of FRP elements to steel joints is the bond of the adhesive [31–33]. The successful implementation of FRP composites of the strengthening systems is dependent upon the quality and integrity of the steel-composite joint and the effectiveness of the epoxy adhesive used [34]. The fatigue performance of the CFRP reinforcement details is necessary to check if, under realistic severe spectrum loading, the adhesive joint performs better than the common fatigue sensitive details found on steel bridge girders, in accordance with Eurocode 3 categories and [35] or American Association of State Highway and Transportation Officials (AASHTO) categories E′ and E [36].

The application of FRP composite for strengthening steel bridges and structural elements has become an efficient option to meet the increased cyclic loads or repair due to corrosion or fatigue cracking. Therefore, the objective of this review is to explore the potential FRP reinforcing techniques to care for fatigue damaged steel structural elements. This paper reviews the research related to CFRP/steel strengthening techniques under fatigue. Detailed knowledge and explanation of the existing research concerning the fatigue behaviour of HM, HS, UHM, and prestressed CFRP and SW-BFRP strengthened steel structures are provided. The study also covers the surface treatment techniques, adhesive curing, and support condition under cyclic loading including fatigue performance, crack propagation, and failure modes with FE simulation of the steel bridge girders and structural elements. Future research gaps and recommendations are indicated accordingly.

SURFACE PREPARATION AND TREATMENT FOR STEEL BEAM STRENGTHENING

Types of Notch

To simulate the actual damage caused by corrosion and the expansion of fatigue cracks, several researchers intentionally created notches of different geometry in midspan or other positions on the tension flange of the beams (Figure 1(a)). In addition, the notch assists like a stress concentrator [37] in the damage-sensitive regions [38, 39] to commence a vertical crack at the steel web. The different types of notch can be categorized as follows:

a. Rectangular notch on both edges (Figure 1(A)),
b. U-shaped notch on both edges (Figure 1(B)),
c. U-shaped notch through the whole tension flange (Figure 1(C)),
d. No uniform notch (Figure 1(D)),
e. Uniform notch (Figure 1(E))

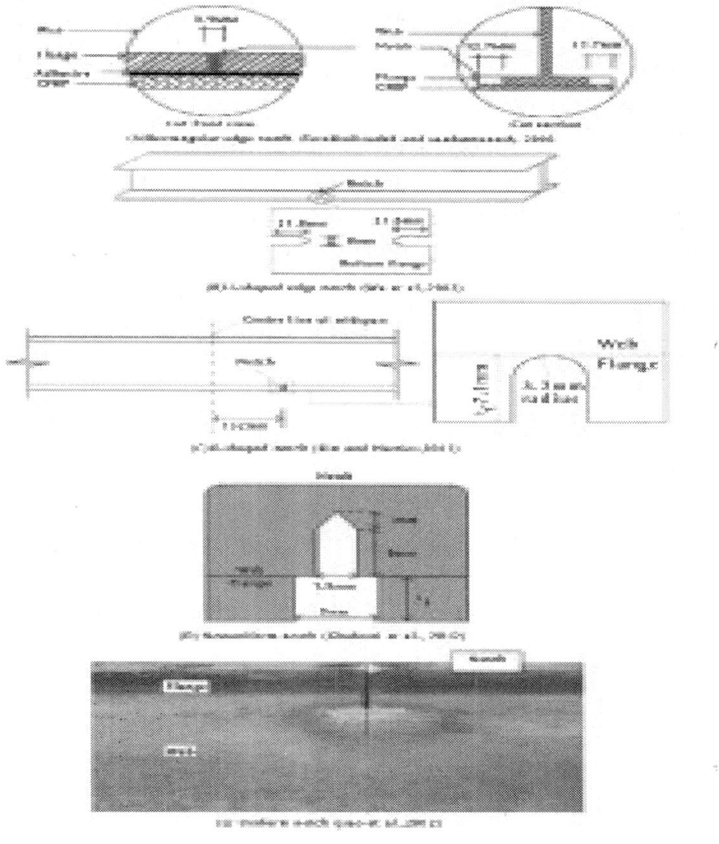

Figure 1: (a) Types of notch in steel beam. (b) Stress behaviour in fatigue for different notch categories [14, 37, and 40].

Usually, the researchers cut a notch at midspan in the tension flange of the steel beams except Kim and Harries [37]. To initiate the debonding of CFRP, which was aimed at propagating towards the right support, they created a notch through the entire tension flange at a position 152 mm to the right of midspan of the beams. Jiao et al. [40] welded the cut along the tension flange soffit using the shielded metal arc welding (SMAW) approach.

The notch (C) through the whole tension flange is sensitive to fatigue compared to the side notches (A) and (B). Accordingly, notches (D) and (E), which go through the whole tension flange with part of the web in midspan, are more sensitive to fatigue damage as the damage occurs suddenly.

From the above categories of creating the notch, it has been revealed that when the notch spreads through the flange as well as the web of the steel beam, the propagation of the cracks exists in the notch line of the flange and the web. Brittle fracture can happen in the case of fatigue. This is injurious for the structural element as no warning is given before failure. In addition, notches only created in the flange can expect a retarding fracture with prior warning. However, if the notch is given in the whole tension flange, there is also the possibility of brittle fracture under heavy repeated load. Therefore, for observing the proper fatigue damage, a rectangular and U-shaped notch on both edges at midspan may be competently incorporated for the development of a standardized test. Figure 1(b) illustrates the stress characteristics of a strengthened steel beam under cyclic loading for incorporation of different notches. The S-N curves show that the uniform notch comprising whole flange and web has the least stress compared to others. This clearly demonstrates more fatigue life when the side rectangular notch is incorporated in the flange

Welding Cover Plate and Stiffener

To provide the equivalent of a concrete slab that typically exists in bridges for preventing compression flange buckling, Wu et al. [38] attached a steel cover plate with welding on the outside surface of the top flange of the steel beams (Figure 2). They also welded web stiffeners on both sides of the web at the loading and supporting points. Web stiffeners assist in preventing web crippling at the midspan section [41]. In order to provide lateral stability of the steel beams, Kim and Brunell [42] used stiffeners that were welded at the supporting points. The adhesively bonded steel stiffeners to the flanges and webs on both sides of the beam extensively retarded local buckling of the steel beams [43]. Siddique and El Damatty [44] showed that the use of glass fibre-reinforced polymer (GFRP) composites enhances local buckling behaviour of wide flange steel beams, which is effective, especially for the case of slender beams. The addition of GFRP plates to the compression flange of a steel beam increases both the load carrying capacity and the deflection at failure. The improvement in the load capacity is independent of the web dimensions of the beams for both plastic and slender beams. The study ignored the use of a stiffener. The mode of failure of the retrofitted slender beams ranges from elastic buckling of the system to GFRP rupture when the thickness of the GFRP is varied from 6.35 mm to 19.0 mm. As the GFRP thickness is significant in their study it could well improve the local bucking.

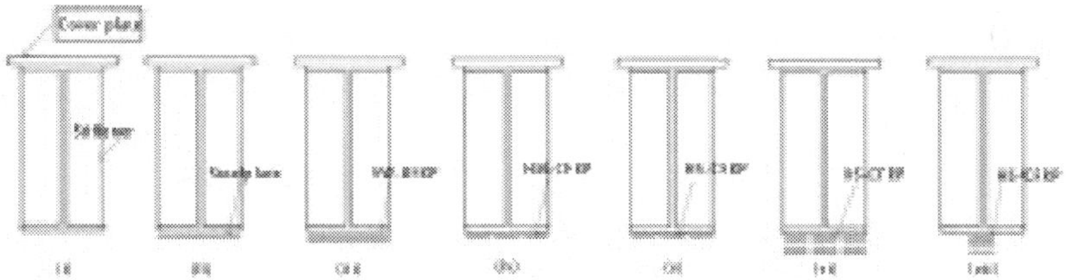

Figure 2: Strengthening technique with plate conf guration [38].

Prevention of Galvanic Corrosion

Although CFRP is a noncorrosive substance, when carbon fibres are in contact with steel, they can form a galvanic cell. To increase the fatigue strength of bridge girders and long-term durabil ty of CFRP reinforcement in a steel structure, the prevention of galvanic corrosion is necessary. Furthermore, to rule out the galvanic corrosion, the flow of corrosion needs to be prevented. This may be accomplished by insulating the different metals from one another or through preventing a continuous link of electrolytic solution between the two by coating with a water resistant sealant [45]. It is obvious that if the two different metals are not in contact, galvanic corrosion cannot occur [46]. Tavakkolizadeh and Saadatmanesh [47] examined the galvanic corrosion between carbon and steel for various thicknesses of adhesive coating in different electrolytes, such as seawater and deicing salt solution. The thin coating effect of adhesive (0.25 mm) was found to be substantial as was the sizing applied to the fibres. Furthermore, a thicker adhesive between the surfaces of the CFRP and steel was observed to suggestively slow the rate of corrosion of the steel.

Mitigating galvanic corrosion of the CFRP-steel composite can be achieved by the selection of an adhesive with good quality isolation properties [15] or by using a thicker epoxy, water resistant sealant, or nonconductive layer plus a sealant, or by bonding a GFRP sheet before applying the CFRP layer onto the steel surface [17, 48–50]. Hollaway and Cadei [51] installed a polyester drape veil to provide insulation between the steel and the carbon fibre for preventing direct contact between them. Fibreglass or an epoxy film was considered to provide effective insulation. In addition, a monitoring programme was initiated to identify the cathodic sites so that galvanic corrosion damage could be mitigated or stopped [34].

Surface Treatment

The reliability of the joint is highly dependent upon the surface treatment processes for bonding the fibre-reinforced composite to the steel structural elements [52]. The surface preparation and the strength of the applied CFRP overlay can significantly affect the fatigue performance [53]. For assessing the effect of the CFRP strengthening technique on the fatigue strength, Jiao et al. [40] used a grinder to remove the corrosion as well as to level the weld area on each steel beam soffit before applying the adhesive. To obtain a clean, rough, and chemically active surface, Wu et al. [38] treated the surface of the tension flange using a grinding wheel to reinforce with CFRP for the fatigue test. The surfaces of the tension flange and composite plates were then washed with acetone.

Tavakkolizadeh and Saadatmanesh [14] used a sand blaster meticulously by number 50 glass bids and washed with saline solution just prior to the application of the composite sheet to prevent oxidation. The study by Teng et al. [54] showed that sand blasting was the most effective surface treatment. Prior to bonding the CFRP strips to the beams, Kim and Harries [37] used a 1500 sfpm (surface feet per minute) belt sander and a 40 grit zirconia alumina belt. This ensured a sound, slightly striated surface to bond the CFRP strips. The adhesive layer thickness was approximately 1 mm.

Deng and Lee [55, 56] found that the tips of the FRP plates must be finished smoothly using sandpaper before the attachment of the plate to the steel beams. However Schnerch et al. [57] disagreed with Choudhury [58], as they contended that preparing the surface with a hand grinder followed by sandpapering reduces the bonding ability of the surface. However, a chemically active steel surface that is free from contaminants is essential to enhance the chemical bond between the adhesive and the metallic surface. Brushing, ultrasonic, or vapour degreasing systems are claimed to be the most efficient to remove oil and other potential surface contamination, especially when adequate solvents are used [59]. Contamination may then be removed using the excess solvent, rather than simply redepositing it on the steel surface as the solvent evaporates.

The most efficient means of achieving a high-energy surface of the steel is by grit blasting [51]. Grit blasting with angular grit removes the inactive oxide and hydroxide deposits by cutting and deformation of the base material. The grit size also affects the surface profile of the steel. Harris and Beevers [60] stated that finer particles created a smoother surface than coarser grit particles and smoother surfaces exhibited higher adhesive-steel surface bonding. In addition, the surface profile of the steel was not influenced on the long-term durability [34]. After grit blasting, solvents may be used to wash and clean the steel surface without resulting in poor bonding [61, 62].

FATIGUE STRENGTHENING TECHNIQUES FOR STEEL BEAMS

The modulus of elasticity, tensile strength, shape, and configuration of FRP composites of an adhesively bonded joint play an important role in respect of the fatigue strength and lifetime of reinforced steel beams and bridge girders. A number of researchers have investigated reinforced steel beams with different FRP strengthening techniques and compared their fatigue performance. A summary of the reinforcing technique of steel structures using fibre-reinforced composites is provided in Table 1.

Table 1: Strengthening techniques

Specimens reference	Material-dimensions and properties	Figures
S127x4.5 steel beams reinforced with CFRP sheet [14].	Steel beam: 1.22 m long, MPa,E=194.4 330.9 MPa, GPa, CFRP: 300 mm × 76 mm × 1.27 mm,E=144 GPa, MPa Epoxy: 1 : 1 mixer of resin (bisphenol A based) and hardener (polyethylene polyamine).	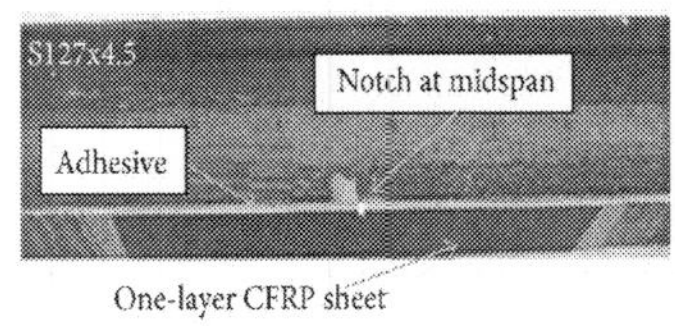
127x76UB13 steel beams strengthened using CFRP plate [55].	Steel beam: 1200 mm long,E=200 GPa, MPa,f_u=275 FRP: 3 mm thick and 400 mm longEpoxy (Sikadur 31 Normal):E=8 GPa, Shear moudulus=2.6GPa,f_u= 29.7 MPa and 0.3 mm thick	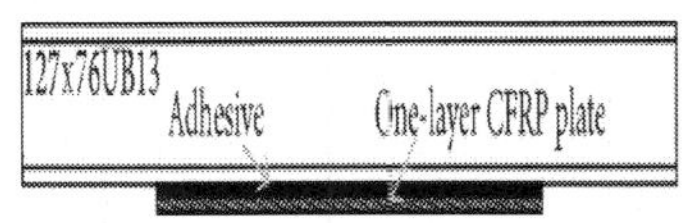
W150x18 steel beams repaired with CFRP strips [37].	Steel beam: span length = 1830 mm,E=200 GPa,f_y=393 MPa CFRP: 50 × 1.4 mm, GPa,f_u= 2.8 GPa Epoxy: GPa,f_u= 25 MPa, 1 mm thick	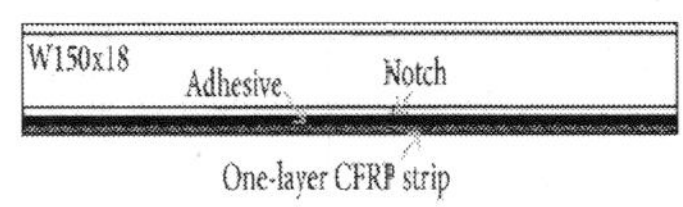
Grade 400 150UB14 notched steel beams retrofitted with welding, CFRP composites [40].	Steel beam: L1400 ×W75×H150 mm,E=207.4 GPa, fy=411.6MPa, f_u=541.3 MPa CFRP:(1) Sika CarboDur M1214 pultruded plates-1.4 mm thick,E=210 GPa, Ft = 2.4 GPa,(2) SikaWrap Hex-230C woven sheets-0.13 mm for each ply,E=230 GPa,=3.45 GPa, Epoxy: (1) Sikadur-330 (2) AralditeR 420	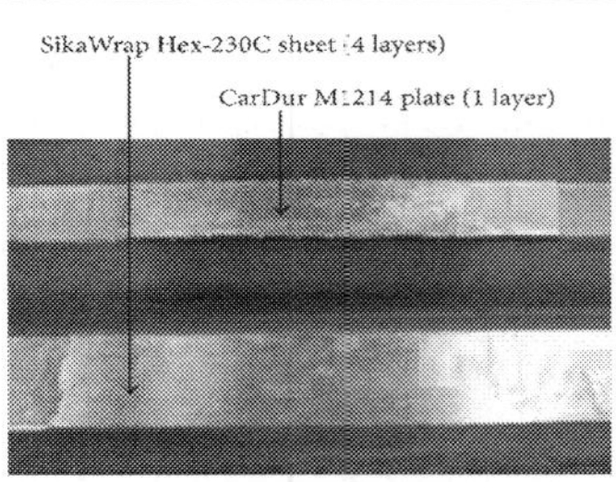
S355J0 steel beams bonded with 20% prestressed CFRP plates [63].	Steel beam: 1100 × 120 × 65 mm, E=210GPa, MPaCFRP: 910 × 50 × 1.2 mm (S512), GPa f_u=3.10 GPa, Prestressing level 20% of the ultimate CFRP strength=632 MPa. Araldite 2015 adhesive:E=1.75GPa,	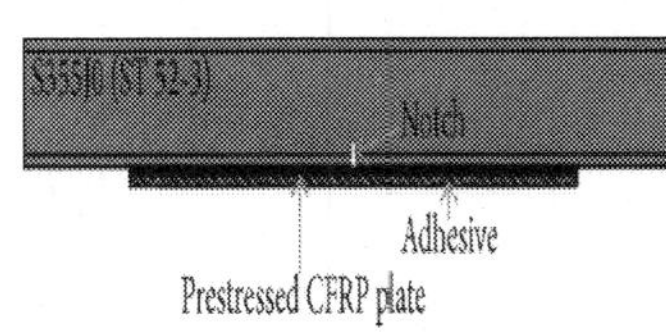

S355J0 (ST 52-3) steel I-beams strengthened using 30% (a) prestressed unbounded and (b) bonded CFRP plates [64].	Steel beam: 11C0 × 120 × 65 mm,E=210CPa,f_y= 355 MPa CFRP: 910 × 50 × 1.2 mm (S512), GPa,f_u= 3.10 GPa, Prestressing level 30% of the ultimate CFRP strength. Araldite2015adhesive: E=1.75GPa,	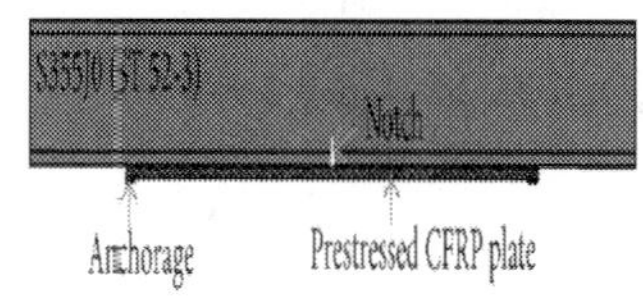 (a) PUR system 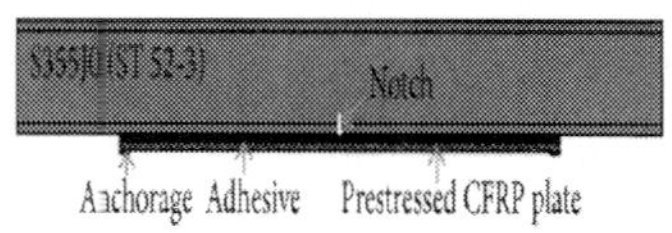 (b) PBR system

Repaired with the welding method alone. It was observed that the strengthening system Furthermore, Wu et al. [38] investigated eight artificially damaged H350 × 175 steel beams, including one unstrengthened and seven strengthened with welded steel, SW-BFRP, high modulus CFRP (HM-CFRP), and high strength CFRP (HS-CFRP) plates using Sikadur-30 Normal epoxy adhesive. The plate configuration of the strengthening technique used by Wu et al. [38] is shown in Figure 2. The elastic modulus, ultimate strain, tensile strength, and shear strength of the epoxy were 2.627 GPa, 1.5%, 31.7 MPa, and 14.4 MPa, respectively. An anchorage system was applied at below the point load and at the end of the fibre-reinforced composite plates. The HM-CFRP has the most excellent strengthening effects and SW-BFRP is the best strengthening material on the basis of the cost-performance ratio [38].

Basalt-FRP (BFRP) composites show synthetical advantages in structural strengthening, seismic rehabilitation, and serving as new structural materials [26, 65]. However, the relatively low modulus of BFRP may not satisfy the stiffness requirement of some structures. Therefore, to obtain higher performance, steel-wire- (SW-) BFRP can be made from hybridization of BFRP with steel wires (SW) or CFRPs [26, 66]. A 210 ft × 26 ft three continuous span rolled steel bridge girder in Guthrie County, Iowa, on state highway 141 was strengthened using externally posttensioning CFRP rods [67]. The anchorage systems were bolted to the webs of the steel girders. The proposed prestressed unbounded reinforcement (PUR) system [64] can be applied as an alternative to adhesively bonded FRP reinforcement, mainly when there is concern about the effects of water, moisture, high ambient temperatures, and high cycle loading on the glue between the FRP and the steel. Vatandoost used 14%, 15%, 35%, 0%, and 37% prestressed CFRP plates to investigate the fatigue behaviour of five W310 × 74 steel beams Vatandoost [68], in which the 14%, 15%, and 35% prestressed CFRP plates were bonded to the inner side of the bottom tension flange, and the 0% and 37% prestressed plates were attached to the cover plate. A prestress FRP composite patch is strongly suggested to maximize the effectiveness of the adhesively bonded patch on the steel element [69] and fatigue strengthening [70]. The FRP prestressing system developed at Empa was used for the CFRP strips for applying a direct tensile force against an external reaction steel frame by jacking. Vatandoost and others [68, 71–73] discussed more details concerning the prestressing procedure.

Recently, the carbon-flex, that is, carbon fibre hybrid-polymeric matrix composite (CHMC) strengthening technique was developed by Zhou and Attard [74], which is a carbon fibre-based composite manufactured using the latest hybrid-matrix technique involving amino-based polymeric composites to provide necessary damping and high strength sustainability of the carbon fibrous element. Recently, Zhou and others [74–76] indicated the enormous potential of carbon-flex as a strengthening substance to subsequently prevent higher damage or catastrophic failure of structures. The majority of fatigue problems arise from poor detailing or careless fabrication, rather than inaccurate materials selection [77]. Schnerch and others [15, 52] reported that the bonding mechanisms of FRP strengthened steel structures are different than concrete structures. In addition, high bonding stresses occur in steel structures to meet the strengthening requirements [34]. Any violation of fabrication tolerances can unpredictably change the fatigue behaviour and lead to a very scattered fatigue life [73].

ADHESIVE CURING

If a bridge or long span structure is retrofitted with CFRP strips, it is generally not economic to close it to traffic during the adhesive curing time, which can take up to 48 hours. During this time, the epoxy adhesive is subject to repeated loading from the traffic. The Concrete Society [78] recommended that the change in the epoxy properties caused by the repeated load during the curing time is expected to be small, perhaps a 10% decrease in the strength of the fully cured structural elements. Nikouka et al. [79] studied the improvement in the strength and stiffness of strengthened steel beams with CFRP subjected to repeated loading during the early age curing of the epoxy adhesive. Five pairs of 127 × 76UB13 type steel beams, each 1.2 m long, were strengthened with a 0.98 m long single K13710 ultra-high modulus CFRP plate, attached to the bottom tension flange. A cyclic load was applied to the five specimens with 0.25 Hz frequency and was continued for up to around 48 hours. The study reported that during the curing of the epoxy, the cyclic loading would affect the final stiffness and failure load of the strengthened beam when the highest cyclic load was larger than 42 kN. Moreover, the bond would fail to develop if the shear deformation in the epoxy layer during the cure is too large. They also recommended that it was prudent to limit the shear stress in the epoxy to a maximum of 1 MPa.

Bourban et al. [80] indicated a clear advantage from the epoxy adhesive curing at high temperatures (about 93°C) during the initial cure (10–20 minutes). The resulting bond is stronger, tougher, and more durable when subject to unfavourable environments [81]. With the intention of retrofitting steel bridges open to traffic during the adhesive curing period, Moy [27] investigated the effect of repeated loading on the curing of the epoxy. The results confirmed a progressive stiffness increase of the reinforced component as the epoxy cured. Furthermore, the beams subject to higher loads during curing did not develop the full bond of the epoxy adhesive. The tests performed showed that cyclic loading at higher load levels reduced the ultimate capacity of the strengthened beams [41]. In addition, the vibration of the traffic during the curing of the adhesive causes a progressive reduction in the fatigue lifetime with increasing strain level [82, 83].

Zhang et al. [84] proposed an innovative method involving preimpregnation (prepreg) advanced composites and a compatible epoxy film for retrofitting steel railway bridges open to traffic during the curing period of the epoxy adhesive. The strengthening system was made from unidirectional HM- and UHM-CFRP preimpregnations that were cured on site under vacuum assisted pressure. Two cure temperatures were used: 65°C for 16 hours and 80°C for 4 hours. A GFRP prepreg layer was placed between the CFRP patch and the steel element. The beams were initially induced by vibration forces and then loaded to failure. From the experimental results it was observed that despite slight damage at the adhesive level, the proposed technique prevented severe brittle failure of the composite beam.

SUPPORT CONDITION AND INSTRUMENTATION

Different support conditions have been adopted by different researchers for the fatigue test programmes of steel beams, as shown in Figure 3. Deng and Lee [55] tested the fatigue strength of nine reinforced steel girders by a servo-hydraulic Dennison testing machine, using a three-point bending setup as a simply supported beam (Figure 3(a)). The specimens were supported on two rollers but were restrained from any sideways movement. The loading block had two steel plates, each with a counter seat and a roller in between. Deflections were measured at three locations by means of potentiometers. Five 2 mm and two 5 mm long strain gauges were used to investigate the crack initiation as well as the effect of crack growth on the stress field in the girder. All the data were recorded using a data logger. Studies on the fatigue of double sided reinforcement subjected to tension and full-scale bridge girders retrofitted with CFRP plates under three-point bending were conducted at the University of Delaware [85]. Kim and Harries [37] used a neoprene rubber pad between the support and beam to reduce the concentration of stress, as shown in Figure 3(b). In all cases, the CFRP plates remained fully bonded to the steel element. The results suggested excellent fatigue behaviour of the reinforced elements.

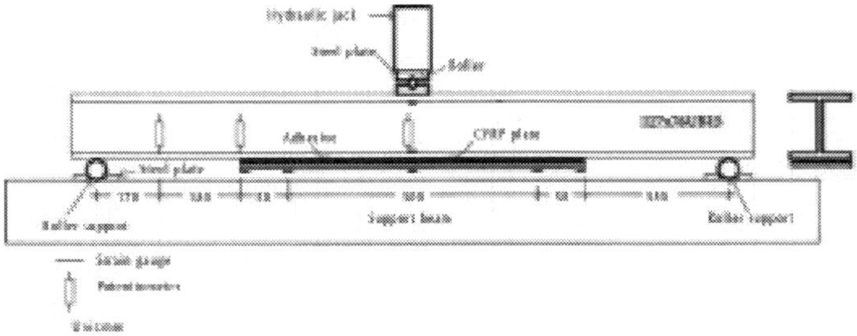

(a) [55]

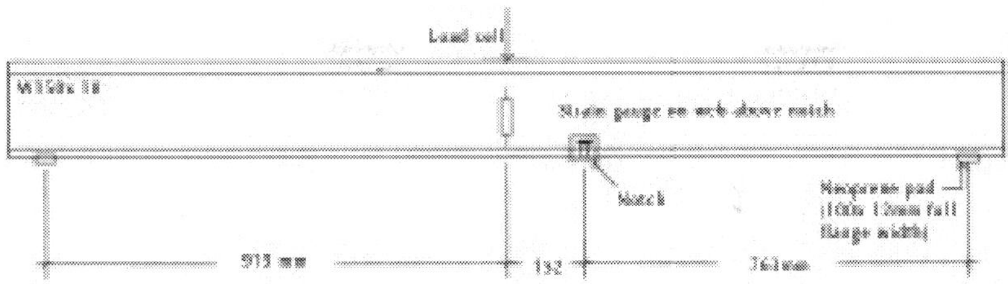
Load cell
Notch

(b) [37]

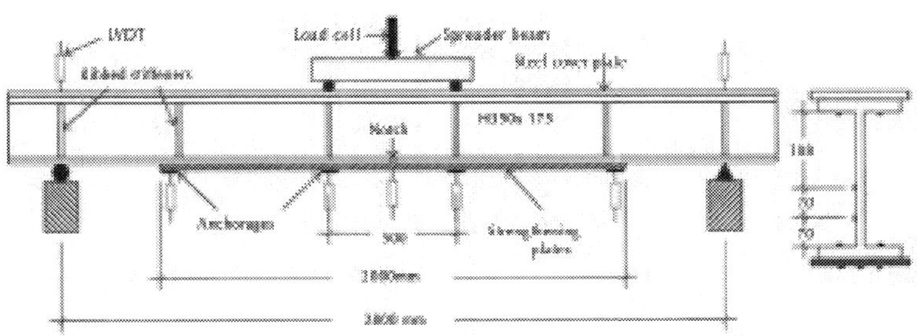
LVDT
Load cell
Spreader beam
Steel cover plate
Notch
Anchorages
Strengthening plates

(c) [38]

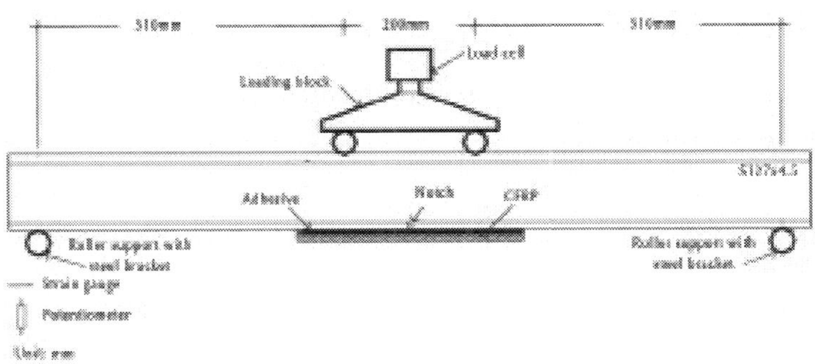
Load cell
Loading block
Adhesive
Notch
CFRP

(d) [14]

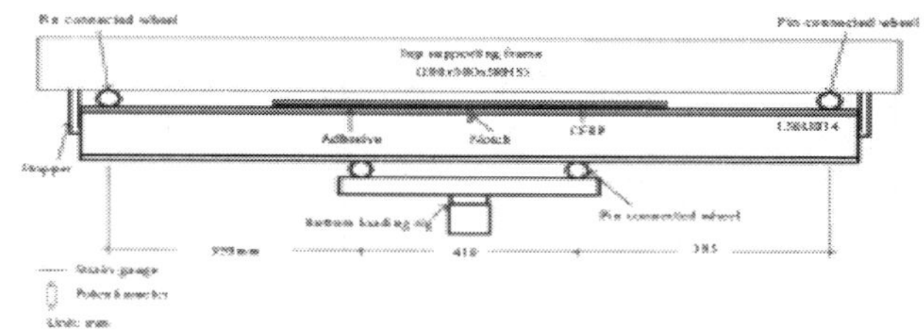

(e) [40]

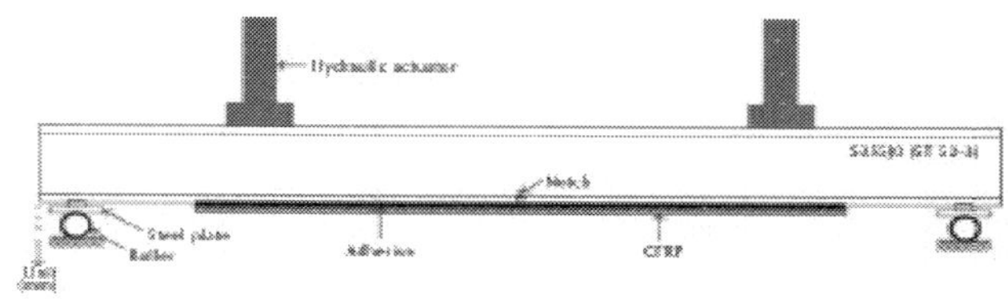

(f) [40, 63]

Figure 3: Support condition with fatigue test setup.

Wu et al. [38] tested the strengthened H350 × 175 steel beams under constant amplitude cyclic load using 4 Hz frequency as a simply supported mode and four-point bending, as shown in Figure 3(c). The load was measured by the loading cell of a MTS system. To prevent any movement of the specimen during the test, Tavakkolizadeh and Saadatmanesh [14] used tie down brackets to the roller supports. The loading blocks were designed using a counter seat for the compression flange in order to prevent their movement during the experiments. The loading setup is shown in Figure 3(d). The specimens were tested using various constant stresses ranging between 69 and 379 MPa and a frequency of 5 and 10 Hz. Vertical displacements can be measured by linear variable displacement transducers (LVDT) with a range of ±50 mm [38] and ±75 mm [14].

Jiao et al. [40] conducted fatigue tests under load control with 7 Hz on strengthened steel beams with a 4-point bending rig using a MTS-

810 testing machine, which contained the top supporting frame and bottom loading beam as shown in Figure 3(e). Two 12 mm thick steel plates were welded to the midspan of the top supporting frame and the bottom loading beam. Four pin connected wheels were employed at the supporting and loading points that could freely rotate during the fatigue tests. Two screw-fixed stoppers were used on both sides of the bending rig to prevent the test specimen from changing position during the experiment. The bottom loading rig was designed using a three-pin system to ensure that the load was distributed between the two loading points. Using SHOWA strip strain gauges, the ultimate load, displacement, number of cycles, and corresponding strains of each cycle were recorded using the National Instrument NI 9237 Compact Data Acquisition system.

To fatigue test the reinforced metallic beams using prestressed FRP, Ghafoori et al. [64] used a pulsator P960 oil hydraulic test machine with a four-point bending setup. The lubricated rollers of 5 cm diameter at the supports and a steel plate were employed between the beam and rollers to distribute the load properly (Figure 3(f)). Ghafoori et al. [63] used a 3D ICS (image correlation system) to measure the crack length and the corresponding strain deformation at the crack tip area. The measurement window of the ICS was set at 65 mm × 65 mm. The calibration details and the use of the ICS can be found in [87–89]. The field signature method (FSM) is also effective for detecting and monitoring cracks on steel structures [90].

FINITE ELEMENT (FE) SIMULATION

The finite element method (FEM) is an acceptable approach for analysing structures using software. In practice, the FE simulation is developed to validate the fatigue strength of the experimental or analytical results.

Based on the surface crack widening energy release rate [91] using an elementary material strength theory [92] and -integral [93], an analytical approach was introduced by Ghafoori and Motavalli [87] to estimate the stress intensity factors (SIF) of a cracked steel I-beam. The fatigue rehabilitation of steel structures is usually expected to decrease the value of SIF at the tip of the crack and, as a result, enhance the postcrack fatigue life [94]. Ghafoori et al. [64] proposed an analytical method using the experimental test data (the external bending moment, the length of the crack, and the corresponding strain imposed on the CFRP strip under the cracked segment) and produced the SIF. They used ABAQUS software (version 6.8) to analyse the FE model of the steel beams to validate the results. The geometrical model and more mesh refinement around the loading, anchorage, and cracked sections are shown in Figure 4. The method was developed to assess the sufficient level of the CFRP prestressing to arrest the fatigue crack growth (FCG). Moreover, the method was used to study different active, semiactive, and passive crack modes with a loaded reinforced beam. Several factors have been considered including crack propagation, excitation frequency, and structural damping on the life of the FCG [95].

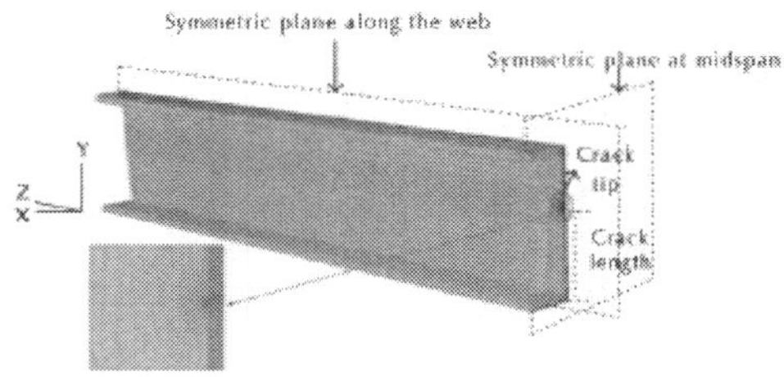

(a)

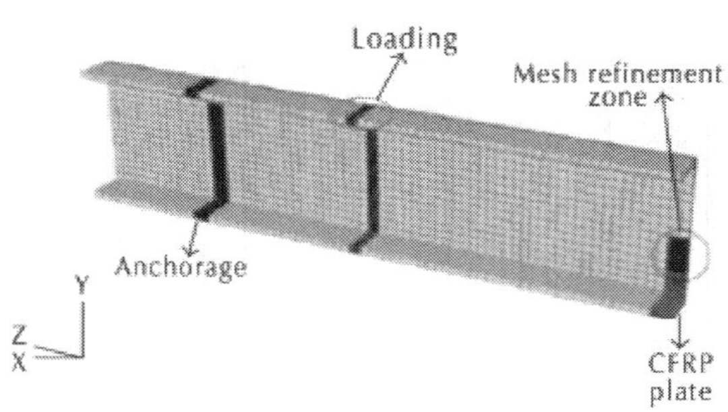

(b)

Figure 4: (a) A geometrical model using ABAQUS in the FE analysis and (b) the mesh refinement around the loading, anchorage, and crack zones [64].

Using the concept of fracture, the fatigue crack propagation (FCP) model was proposed by Xiulin and Hirt [96]. This was extended to the FCP of a cracked metallic element retrofitted with adhesively bonded composite patches in the study of Wang and Nussbaumer [97].

According to the Paris-Erdogan crack growth law [98], a linear elastic fracture mechanics (LEFM) model was employed to predict the effects of peening treatments on the fatigue performance of welded steel structures [99] and to confirm the effectiveness of the prestressed CFRP strips [68]. Ghafoori et al. [63] introduced a methodology for a damaged beam with a specific crack length that is subjected to a certain cyclic load based on the fracture mechanics (FM) theory to estimate the adequate prestressing level by which the crack propagation is detained. Some strengthened beams were tested under various cyclic loading ranges and the experimental results showed excellent agreement with the developed fracture model.

In fracture mechanics, based on progressive damage modelling methods to predict the fatigue life, the rate of crack growth is related to the SIF [98] or strain energy release rate [100–103]. In the case of adhesively bonded joints, a damage shift parameter was proposed to account for the effects of accelerative interaction [102, 103]. Important interaction effects were considered where crack growth acceleration was linked with mean load changes. However, continuum damage mechanics (CDM) models were developed by Lemaitre and Desmorat [104] and modified for the damage formation of microcracks by [105, 106]. CDM models have been used in a damage evolution law for modelling both precracking damage evolution and crack growth for constant and variable amplitude fatigue [107]. In the case of bonded joints, Wahab et al. [108] compared both the FM and the DM methods to predict the fatigue strength of adhesively bonded CFRP double lap joints. They verified that the developed CDM approach compared favourably with a FM method for constant amplitude fatigue (CAF). The FM and DM based fatigue life prediction of bonded single lap joints (SLJs) subjected to different types of variable amplitude fatigue (VAF) loading was analysed by Shenoy et al. [109].

Kim and Harries [37] developed a three-dimensional (3D) nonlinear finite element model for predicting the fatigue strength of notched steel beams using ANSYS software. The steel section was modelled using 3D structural solid elements (SOLID45); and a linear stress-strain relationship was developed for the CFRP. A nonlinear interface element (COMBIN39) with two nodes was applied for modelling the behaviour of the steel-CFRP interface. For the element whose initial relative distance is zero, a bilinear bond-slip relationship was created for them. The study used the strain life method and the concept of Henry's damage theory [110] for the fatigue life prediction of steel beams. The strain life approach is mainly relevant to a member representing significant plasticity induced by hysteretic loads. The theoretical background of this approach is discussed by Bannantine et al. [111]. The deflection behaviour of unstrengthened and strengthened beams is shown in Figure 5(a). Furthermore, a typical S-N curve of strengthened steel beam obtained is shown in Figure 5(b), which was compared with category E in the AISC in the study by [112]. The notch provided for the stress concentrating effect is essentially equivalent to a Category E detail. Apart from this, Youssef [113] developed a model for predicting the linear and nonlinear behaviour including the deflection at midspan, strains of the steel and FRP, failure mechanism, and failure load of rehabilitated steel beams. The model was founded on the solution of the differential equations governing the behaviour of a strengthened steel beam, which includes representation of the shear and peel behaviour of the epoxy adhesive. To validate the predictions of the model, a W-shaped steel beam strengthened using GFRP sheets was experimentally tested and excellent agreement was found between these results.

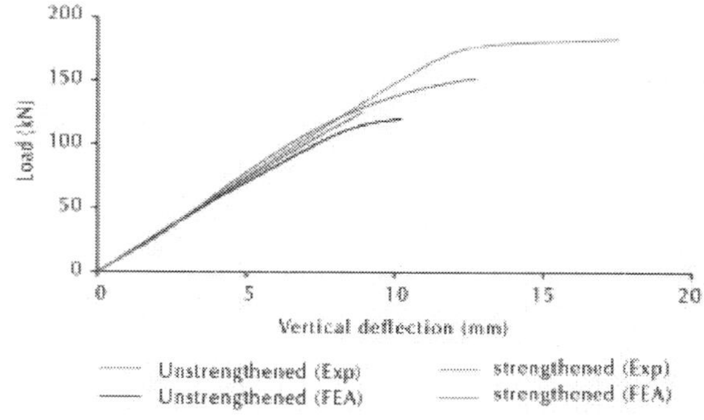

(a)Load deflection relationship in experimental and numerical studies (static) [86]

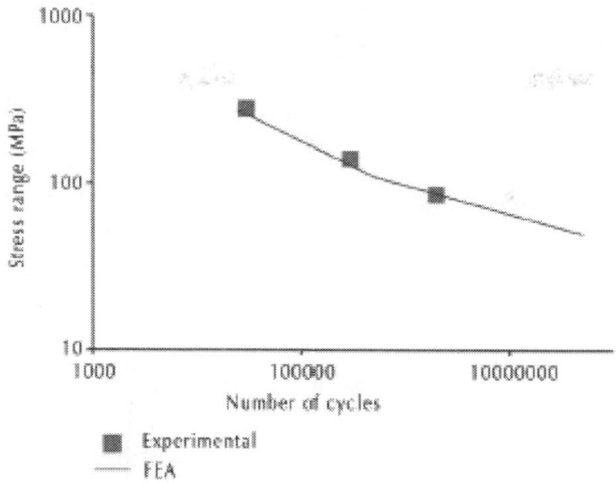

(b) S-N curve in experimental and numerical studies [37]

Figure 5: Comparison of experimental and simulation outputs of steel beam.

Zhou et al. [114] adopted the micromechanics based fracture model and the cyclic void growth model (CVGM) for estimating extremely low cycle fatigue (ELCF) fracture of the column-to-beam connections during earthquakes. The model was verified by the experimental results of nine full-scale connection tests. In addition, the refined finite element model was used to simulate the cyclic behaviour of the connection tests, and the CVGM fracture index was calculated using the stress and strain time histories. The number of cycles and the cumulative deformations to ELCF fracture predicted by CVGM agreed well with the experimental results. The existing methodology also showed reasonable good accuracy for predicting the ELCF fracture of column-to-beam connections under inelastic cyclic loadings.

Pipinato et al. [115] used a LEFM method in a probabilistic [116] context to assess the fatigue reliability of steel bridge girders in the presence of seismic loading. This method could enable a better understanding of progressive damage phenomena due to fatigue problems and could give some new insights to increase the remaining fatigue strength of a large number of steel bridges in seismic regions. Colombi [117] developed a suitable plasticity based [118, 119] crack retarding model as an extension of the well-known Newman's model [120, 121] to estimate the reduction of crack opening displacement along with the magnification of the crack growth retardation of the reinforced notched steel plates.

From the abovementioned literature, it is revealed that the simulation in the finite element method can be a vital tool to assist in strengthening beam analysis under fatigue. This is because it eventually decreases the experimental cost in finance and time. Good validation of the simulation with practical experiments ensures the advantages of the strengthening techniques. However, the characteristics of strengthened steel beams under fatigue without using notches are still an interesting area to be explored. This interest can be addressed by FE simulation in a consistent manner.

FATIGUE PERFORMANCE OF REINFORCED STEEL BEAMS

A steel structure subjected to repeated load may eventually experience significant fatigue damage during its life. A number of researchers concentrated on fatigue strength and fatigue lifetime prediction of reinforced steel beams and bridge girders. This indicates that there is a need to enhance the fatigue strength and prolong the fatigue lifetime of steel structures with adhesively bonded metal/FRP strengthening techniques. The fatigue behaviour of reinforced steel beams using nonprestressed and prestressed FRP composite is illustrated below.

Reinforced With Nonprestressed Polymer Composite

Hollaway and Head [122] indicated that unidirectional continuous fibre polymer composites, which essentially behave linearly up to failure level when loaded parallel to the longitudinal fibres, usually have good fatigue properties. Jiao et al. [40] compared the behaviour of notched steel beams using the welding method and retrofitted with CFRP plates and sheets, respectively, under flexural cyclic loads. In addition, two different epoxy adhesives, that is, Sikadur-330 and Araldite 420, were used in this test. The observations of the fatigue strength of the specimens reinforced with CFRP composites were extensively longer than that of specimens with one layer of CFRP plate adhesively bonded could extend the fatigue strength of steel beams about seven times compared to the beams only repaired with the welding method. In addition, the fatigue strength was extended about three times for beams strengthened with four layers of CFRP woven sheets, and no significant variation in fatigue strength could be observed for specimens strengthened using the epoxy adhesives of Araldite 420 and Sikadur-330. Mean S-N curves were obtained based on the test data (Figure 6), which can be used for predicting the fatigue strength of steel beams strengthened with similar CFRP composite materials.

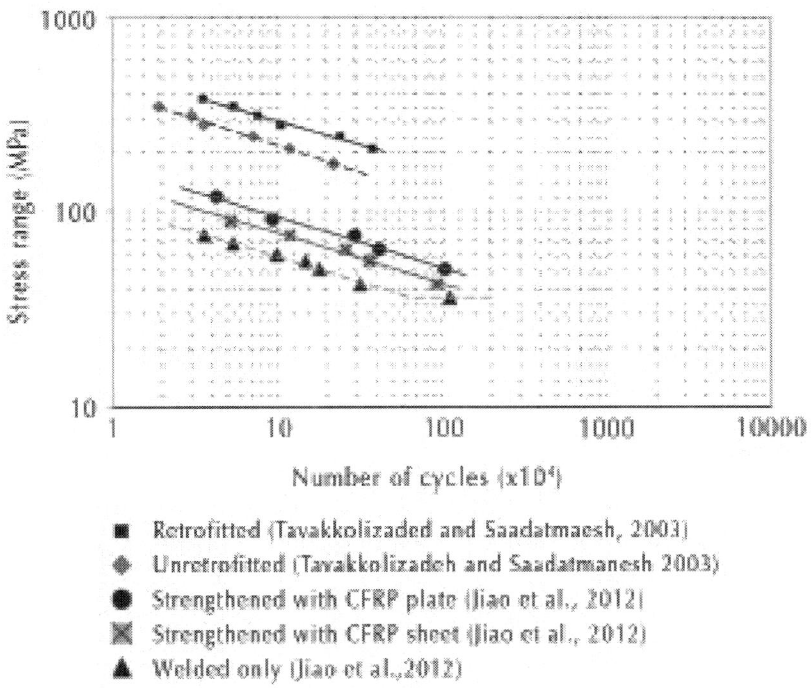

Figure 6: S-N curve for steel beam strengthened with nonprestressed CFRP [14, 40].

Tavakkolizadeh and Saadatmanesh [14] demonstrated using the S-N curve that the application of CFRP strips could increase the fatigue lifetime of the structural elements more than three times (Figure 6). The design S-N curve for unreinforced and reinforced cut specimens was and, respectively.

The slope of the S-N curves for the specimens (retrofitted and unretrofitted) in a log-log space was slightly smaller than the slope of the AASHTO design curves. They also observed that the CFRP patch not only decreased the crack growth rate but also was able to carry a few extra cycles even after the tension flange had completely cracked, especially under lower stress ranges (Figure 7).

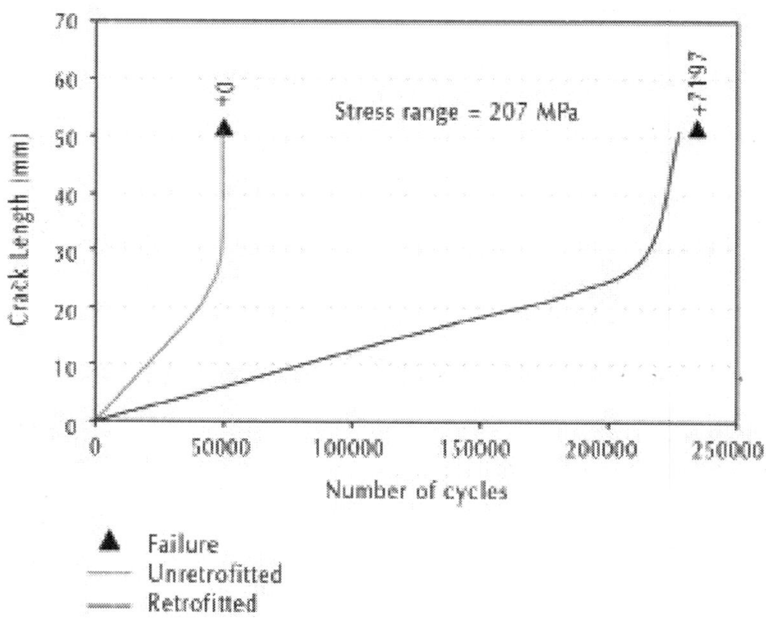

Figure 7: Fatigue crack growth curve for unreinforced and reinforced steel beams [14].

Deng and Lee [55] reported the results of an experimental programme on small-scale steel beams reinforced by applying CFRP strips. From the tests results an S-N curve was obtained. The fatigue limit, that is, threshold of the S-N curve was about 30% of the ultimate static failure stress, which validates the fatigue limit recommended by the CIRIA Design Guidance [123]. To assess the fatigue bond resistance of a steel bridge girder reinforced with CFRP strips, Miller et al. [85] conducted two test programmes. First, they subjected seven small-scale, doubly reinforced specimens to cyclic loads at 82.7 MPa stress for 2.55 million cycles. All CFRP strips were found to remain fully bonded to the steel element without deterioration based on the strain data taken before and after the cyclic loading. Subsequently, two full-scale steel bridge girders retrofitted with CFRP were fatigued for ten million cycles at a constant stress range of 34 MPa. Throughout the tests, the CFRP strips were inspected and monitored for debonding, but none was detected. Therefore, the retrofitting technique was regarded as having good fatigue resistance.

Abed [124] investigated the effects of temperature on the adhesively bonded steel beams reinforced with CFRP composites. The adhesive materials showed a significant reduction in the fatigue life and failure load of the strengthened structures as the temperature reached the adhesive glass transition temperature (T_g). Furthermore, Keller and Schollmayer [125] experimentally and numerically investigated the through-thickness performance of adhesively bonded FRP bridge decks and steel girders. They found that no stiffness degradation occurred for cyclic loading of up to 10 million cycles.

Wu et al. [38] investigated the fatigue behaviour of strengthened artificially notched steel beams including the effects of the configuration and the number of layers of HS-CFRP strips, the interface treatment of the SW-BFRP composites, and the type of materials. Compared to the traditional welded steel-plate approach, the experimental results showed that the use of a fibre-reinforced composite strip could not only delay crack initiation, decrease the crack growth rate, and prolong the fatigue life, but also reduce the residual deflection and stiffness decay. The rough surface of the SW-BFRP could extend the fatigue strength of steel beams more effectively than using SW-BFRP with a smooth surface. They also used high modulus CFRP strips as a reinforced material. HM-CFRP demonstrated the best strengthening performance; the fatigue strength of the steel beams could be improved significantly by increasing the number of layers of the strengthening material. Figure 8 presents the crack expansion curves for four-layer and one-layer HS-CFRP. When the number of layers was increased from one to four, the crack initiation life and the fatigue lives were considerably enhanced (Figure 8). The plate configuration influences the fatigue strength.

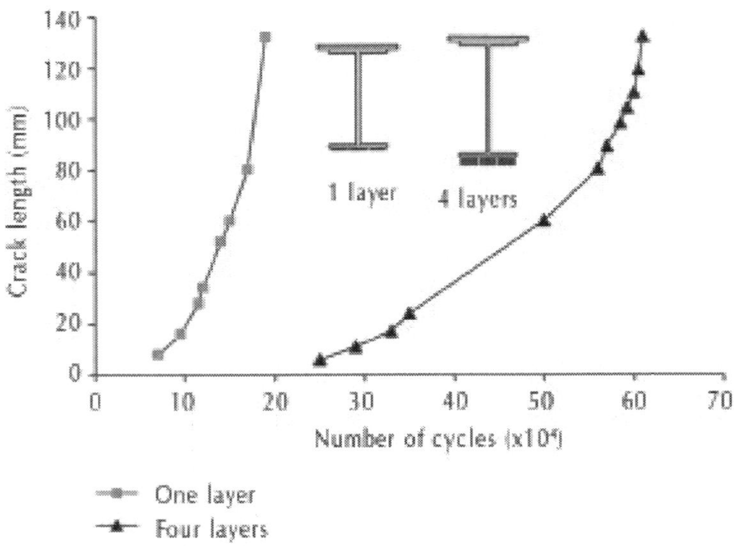

Figure 8: Effect of layers of strengthening material [38].

Reinforced with Prestressed Polymer Composite

Although adhesively bonded FRP flexural strengthening techniques have been found to be an efficient approach to improve the lifetime of fatigued steel structures, there are relatively few studies that have applied prestressed CFRP strips to strengthen against cyclic loadings. Ghafoori et al. [64] developed a prestressed unbounded reinforcement (PUR) method and compared the effectiveness and feasibility of the approach with the prestressed bounded reinforcement (PBR) method. It could be used on heritage and historical structures where reversibility is important. The experimental test results for the strengthened beams using the PBR method showed a local strain concentration on the CFRP strip under the cracked section, while the PUR method had a uniform strain distribution along the CFRP strip. In addition, the fatigue performance of the unbonded reinforcement system was better at a high prestressing level of the CFRP without a substantial reduction in ductility.

Ghafoori et al. [63] studied the behaviour of notched steel beams reinforced using prestressed and without prestressed bonded CFRP plates under cyclic loading. The experimental results showed that the fatigue strength of a beam reinforced using the prestressed CFRP plate increased more than five times that of an identical beam reinforced using nonprestressed CFRP plate (Figure 9). Both specimens were induced by a cyclic load range of 9–90 kN.

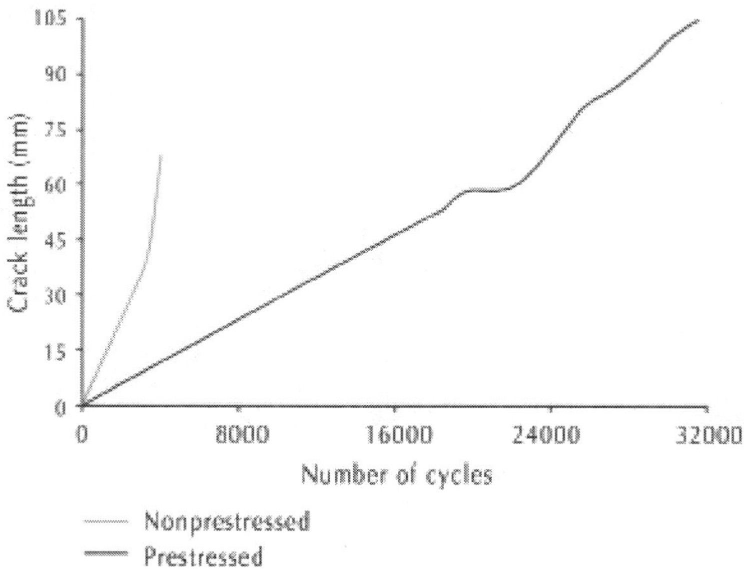

Figure 9: FCG curves for strengthened with nonprestressed and prestressed CFRP plates [63].

Vatandoost [68] drew deflection range curves for W310 × 74 steel beam specimens strengthened with 14%, 15%, 35%, 0%, and 37% prestressed CFRP plates. The deflection range versus the number of cycle curves was drawn for the last 45000 cycles. Looking at Figure 10, it can be seen that (1) the highest deflection range belongs to the control beam while the lowest deflection range belongs to specimen 37%-C-M indicating the highest stiffness increase for that specimen. (2) The deflection ranges are dramatically increased at the end of the fatigue life. (3) The lower deflection range for the specimens with CFRP strips on the cover plates confirms the influence of the CFRP strip location on the specimen stiffness. In Figure 10, "F" indicates that the strips are bonded to the inner side of the flange, "C" indicates that the strips are attached to the cover plate, and "S and M" indicate the CFRP strip with a standard modulus and moderate modulus, respectively. The beams with reinforcement located on the cover plates showed a greater fatigue life improvement than those with reinforcement located on the inner side of the flange [68].

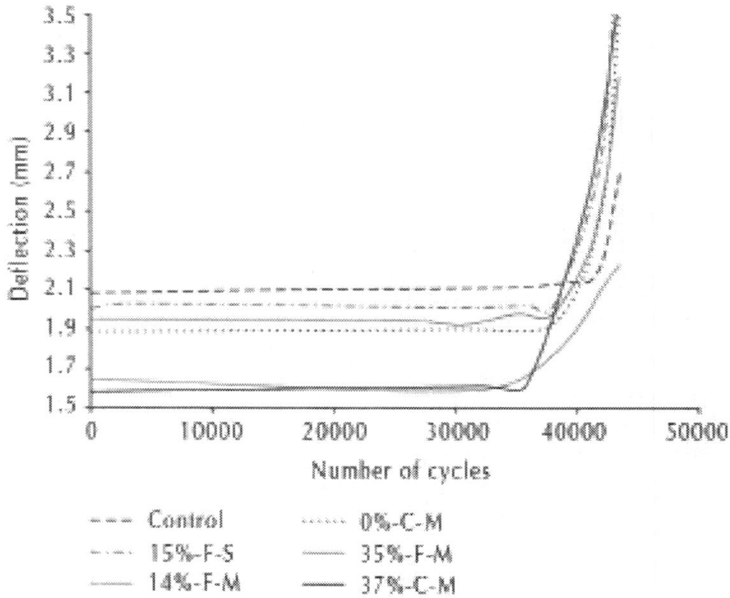

Figure 10: Deflection versus number of cycle curves for strengthened steel beam using 14%, 15%, 35%, 0%, and 37% prestressed CFRP plates [68].

Fatigue in Hybrid-Joint and Nanoadhesive

In recent years, fatigue in hybrid adhesive joints, which combine a traditional mechanical joint and a layer of adhesive (bolted/bonded, welded/bonded, and rivet/adhesive), has attracted a considerable amount of researchers. This is due to their better fatigue performance compared to only mechanical joints or only bonded joints [126–130]. Furthermore, the use of nanoadhesives (carbon nanotubes, alumina nanoparticles, and quartz nanoparticles) is a new field of application to bonded joints and has the potential to improve their fatigue performance [127, 131].

FAILURE MODES OF REINFORCED STEEL BEAMS

The deterioration of the steel bridge structural capacity over time may be due to corrosion, impact damage, and/or fatigue cracking [32, 51, 132]. The crack propagation and failure mode of FRP-strengthened steel structural techniques are normally different than concrete-FRP techniques [15]. Tavakkolizadeh and Saadatmanesh [14] studied the performance of steel I-beams with an edged notch in the tension flange and reinforced with a bonded CFRP patch under cyclic loading. They demonstrated that cracking started from the tip of one of the notches in the reinforced specimens and moved towards the web (Figure 11(a)). In Figure 11(a), the term "near" identifies the side that the crack started and "far" identifies the side that the crack terminated. After reaching the fillet section of the web, debonding started at the near edge of the CFRP patch. While the crack moved towards the far side, the debonding at the edge continued to grow. Even after the crack reached to the far cut end, the debonding remained fairly stable. Then the far edge of the CFRP patch started to debond. The reinforcement failed after around 50 mm of debonding on both sides.

(a)

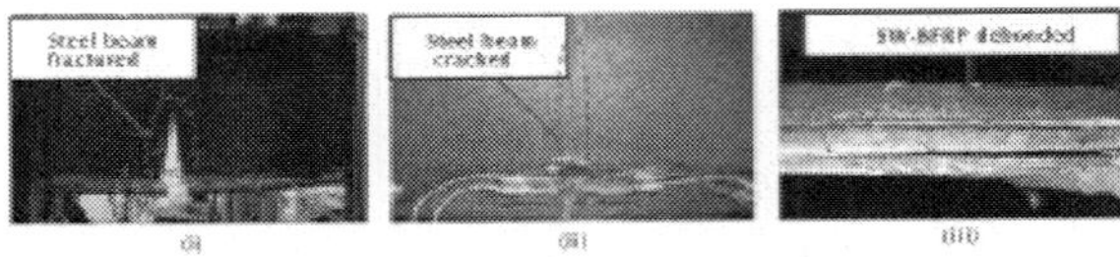

(b)

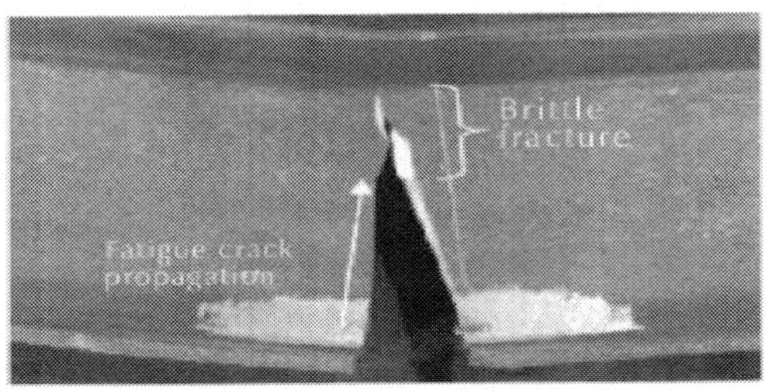

(c)

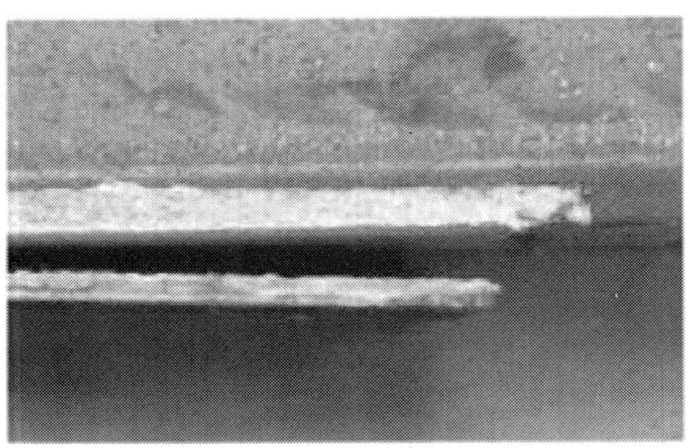

(d)

Figure 11: Failure modes of retrofitted beams using FRP plates (a) [14], (b) [38], (c) crack propagation and brittle web fracture [37], and (d) end debonding [55].

Wu et al. [38] studied the fatigue behaviour of seven steel beams strengthened using four different strengthening materials. During the test, the fatigue crack always started from the notch tip at midspan as the loading cycles increased, mainly because of the stress concentration in this area. Initially, the cracks expanded slowly along the tension flange, and these cracks expanded at an increasing rate as the load cycled. For the steel beam retrofitted by a welded steel plate, the steel beam with plate fractured immediately when the crack moved through the tension flange, as shown in Figure 11(b)(i) and could not bear any further fatigue cycles. For specimens reinforced with FRP composite plates, extra loading cycles could still be sustained when the crack moved through the tension flange. As the load cycled, the crack expanded upwards on the web (Figure 11(b)(ii)) until debonding failure (Figure 11(b)(iii)). No fatigue rupture was sustained by the reinforced composite plate. Hence, the use of a reinforced composite plate can significantly improve the failure mode of the steel beam when compared to the use of a welded steel plate. The difference in failure modes was generally related to the fatigue strength and load-transferring mechanism of the reinforcement material. However, the failure modes depend on the elastic modulus of the FRP and the adhesive thickness [133].

Kim and Harries [37] investigated six beams intentionally damaged by notching the tension flange of the beam to appraise the static and fatigue performance of the beams with emphasis on local plasticity as well as the CFRP-steel interface. The retrofitted beam subjected to a high cyclic loading was observed to exhibit a large brittle fracture in the web (Figure 11(c)). The form of brittle fracture being addressed had been termed "constraint-induced fracture" and could occur without any noticeable fatigue crack growth and, more importantly, without any warning [134].

The FRP reinforcement ends and the regions where geometric discontinuities (cracks) take place are the most sensitive zones to fatigue damage of the adhesive joint, because of the stress concentration [28]. Deng and Lee [55] found similar crack initiation and propagation in the CFRP reinforced steel beams except one test specimen, which was investigated with a 92.6% load range that caused it to debond suddenly at one end after only 30 cycles, Figure 11(d). For the other specimens, the crack at each end grew quickly but then almost stopped after one of the two cracks had advanced past the midspan of the beams. For all the plates that had debonded, the cracking initiated from the middle of the spew fillet and then propagated to the steel surface and glue interface at a 45° angle.

Jiao et al. [40] observed various failure modes for beams strengthened using CFRP plates and sheets. A cut was made and welded along the bottom tension flange of each beam. For unstrengthened beams, a crack was initiated rapidly at the tip of the cut once the crack propagated along the weld on the beam soffit throughout the whole tension flange. It should be mentioned that the beam was considered to have failed when it lost the ability to carry the maximum load applied in every cycle due to severe crack propagation. For strengthened beams with CFRP plates, debonding occurred in the steel surface and epoxy bond interface (Figure 12(a)). Again, failure happened between the plies of the sheets in the strengthened beams as shown in Figure 12(b). The observation also confirmed that the performance of the CFRP plates to resist crack propagation was better than for the CFRP sheets under cyclic loading.

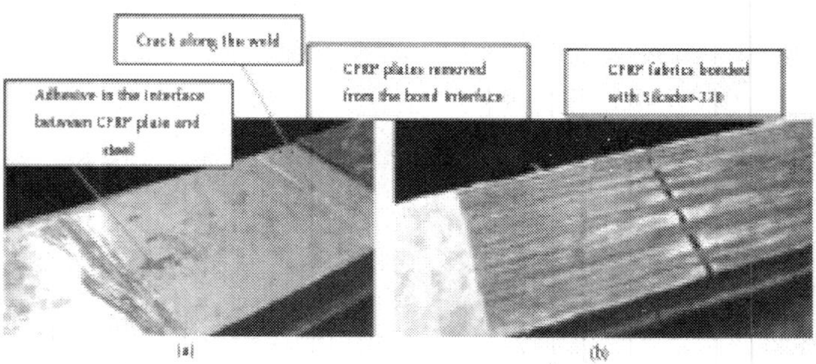

Figure 12: Debonding interface of retrofitted beams using (a) CFRP plates; (b) CFRP sheets [40].

The crack propagation rate depends on the stiffness of the FRP strip and, largely, on the prestressing force [73]. Ghafoori et al. [64] investigated the damaged steel beams reinforced with the PUR and PBR methods and obtained a similar load carrying capacity; however, the failure modes were different. In the PUR method the CFRP plate slipped in the mechanical anchorage system at the onset of the failure, while the CFRP in the PBR method arrived at its tensile strength in the cracked section and, finally, plate failure with debonding (Figure 13).

(a)

(b)

Figure 13: The failure modes for (a) PUR technique and (b) PBR technique.

The end anchoring technique mitigates the debonding of the FRP strips and maintains the prestressing force, hence also reducing the transfer length [68]. Figure 14 illustrates the effectiveness of the end anchoring technique in maintaining the CFRP prestress. The debonded

CFRP strip rupture and the strip end debonding model using FEA is shown in Figure 15.

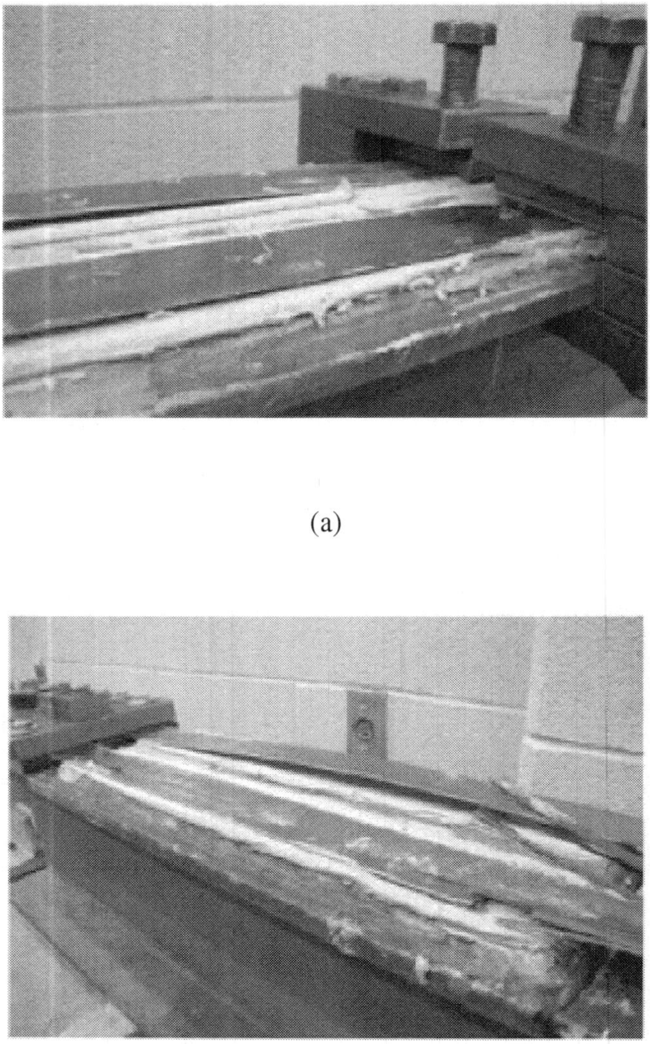

(a)

b)

Figure 14: Performance of end anchorage after failure for steel beam strengthened with prestressed strips to the cover plate (a) removable end anchor and (b) fixed end anchor [68].

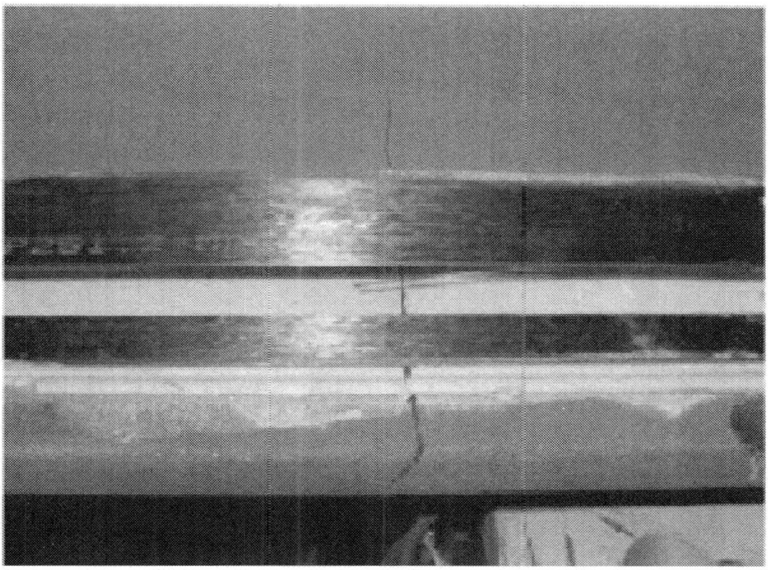

(a)

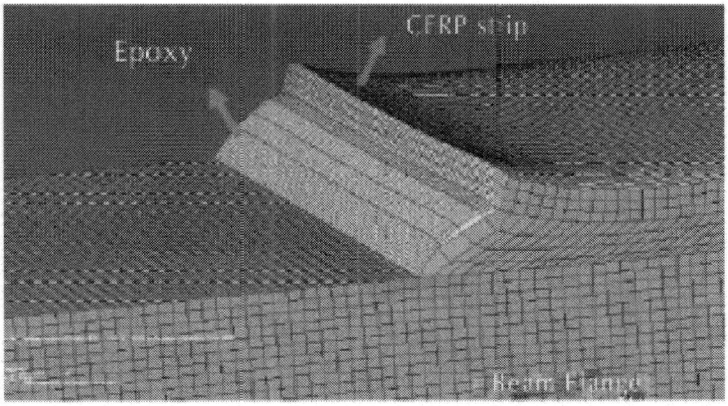

(b)

Figure 15: (a) Debonded strip rupture, (b) end debonding model using FEA [68].

CONCLUSION

In this paper, detailed reviews on the relevant researches were investigated systematically and carried out in respect of CFRP/steel strengthening techniques under fatigue. Significant information and an explanation of the existing research on the fatigue behaviour of FRP-strengthened steel structures have been provided. The study also covered the surface treatment techniques, adhesive curing, and support condition under cyclic loading including fatigue performance, crack propagation, and failure modes with FE simulation of the steel bridge girders and structural elements.

The following conclusions can be drawn from the present study.

1. The remaining service life of bridge structures is limited by fatigue damage, and, in order to ensure the safety of the steel bridges, it is important to regularly check the structure to determine the existence of fatigue cracks.
2. The application of HM, HS, UHM, and prestressed CFRP and SW-BFRP strengthening composites not only delays the initial crack, reduces the crack growth rate, and extends the fatigue life, but also decreases the stiffness decay with residual deflection. In addition, strengthened with prestressed CFRP had the best strengthening effect.
3. The use of end anchorage prevented debonding of the CRRP strips at the beam ends by reducing the local interfacial shear and peel stresses.
4. Epoxy adhesive curing is needed for potential FRP strengthened structures. Cyclic loading during adhesive curing can decrease the fatigue life of the reinforced beam by reducing the bond strength.
5. Alteration in failure modes is mainly related to the fatigue strength and load-transferring mechanism of the strengthening material.
6. The prestressing force should be released very carefully to avoid debonding of the strengthened steel beams adhesively bonded with prestressed FRP.
7. Hybrid-joint, nanoadhesive, and carbon-flex can also be attractive for strengthening systems.

The following recommendations are suggested for future research in this area.

1. An effective surface treatment technique needs to be developed for use in practice that can prevent failure at the adhesive/steel interface under cyclic loading.
2. A comparative analysis of the fatigue behaviour of reinforced steel beams supported on a neoprene pad and steel plate in a simply supported manner should be investigated.
3. The effects of the thickness of the adhesive on the system require more investigation. In addition, the optimum adhesive thickness should be studied to slow down delamination.
4. The length of optimum reinforcement should be determined to retard fatigue failure like end debonding.
5. The influence of different environmental conditions must be studied to find the real fatigue behaviour of the strengthening scheme.
6. The effect of an acidic or alkaline environment on the fatigue strength has not been studied yet.
7. The fatigue behaviour of damaged steel structural elements reinforced with carbon-flex strengthening techniques should be investigated to enhance the cyclic load bearing capacity.
8. An appropriate rehabilitation method using FRP composites of welded steel structural elements under cyclic loading should be explored to retrofit welded steel bridge girders.
9. The application of nanoadhesives to FRP/steel bonded joints for rehabilitation of steel bridge girders to increase the fatigue life has not been investigated yet.
10. The use of hybrid-joints to strengthen steel structural elements with FRP requires further study.

ACKNOWLEDGMENT

The authors gratefully acknowledge the support given by University of Malaya (UM) for funding the work through the High Impact Research Grant H-16001-00-D000036.

REFERENCES

1. J. Bień, L. Elfgren, and J. Olofsson, Sustainable Bridges: Assessment for Future Traffic Demands and Longer Lives, Wydawnictwo Edukacyjne, Dolnośląskie, Poland, 2007.
2. E. Ghafoori and M. Asghari, "Dynamic analysis of laminated composite plates traversed by a moving mass based on a first-order theory," Composite Structures, vol. 92, no. 8, pp. 1865–1876, 2010.
3. Y. Kondo and K. Okuya, "The effect of seismic loading on the fatigue strength of welded joints," Materials Science and Engineering A, vol. 468–470, pp. 223–229, 2007
4. Y. Lu, L. Mao, and P. Woodward, "Frequency characteristics of railway bridge response to moving trains with consideration of train mass," Engineering Structures, vol. 42, pp. 9–22, 2012.
5. S. Hao, "I-35W bridge collapse," Journal of Bridge Engineering, vol. 15, no. 5, pp. 608–614, 2010
6. K. Nishikawa, J. Murakoshi, and T. Matsuki, "Study on the fatigue of steel highway bridges in Japan," Construction and Building Materials, vol. 12, no. 2–4, pp. 133–141, 1998.
7. Y. J. Kim and P. J. Heffernan, "Fatigue behavior of externally strengthened concrete beams with fiber-reinforced polymers: state of the art," Journal of Composites for Construction, vol. 12, no. 3, pp. 246–256, 2008.
8. L. Van Den Einde, L. Zhao, and F. Seible, "Use of FRP composites in civil structural applications," Construction and Building Materials, vol. 17, no. 6-7, pp. 389–403, 2003.
9. N. G. Tsouvalis, L. S. Mirisiotis, and D. N. Dimou, "Experimental and numerical study of the fatigue behaviour of composite patch reinforced cracked steel plates," International Journal of Fatigue, vol. 31, no. 10, pp. 1613–1627, 2009.
10. H. Liu, R. Al-Mahaidi, and X.-L. Zhao, "Experimental study of fatigue crack growth behaviour in adhesively reinforced steel structures," Composite Structures, vol. 90, no. 1, pp. 12–20, 2009.

11. Q. Q. Yu, T. Chen, X. L. Gu, X. L. Zhao, and Z. G. Xiao, "Fatigue behaviour of CFRP strengthened steel plates with different degrees of damage," Thin-Walled Structures, vol. 69, pp. 10–17, 2013.
12. A. Shaat, D. Schnerch, A. Fam, and S. Rizkalla, "Retrofit of steel structures using fiber-reinforced polymers (FRP): state-of-the-art," in Proceedings of the Transportation Research Board (TRB) Annual Meeting, CD-ROM (04-4063), 2004.
13. K. A. Harries, A. Varma, S. El-Tawil, et al., "Steel-FRP composite structural systems," in Proceedings of the 6th International Conference on Composite Construction in Steel and Concrete, pp. 703–716, 2011.
14. M. Tavakkolizadeh and H. Saadatmanesh, "Fatigue strength of steel girders strengthened with carbon fiber reinforced polymer patch," Journal of Structural Engineering, vol. 129, no. 2, pp. 186–196, 2003.
15. X.-L. Zhao and L. Zhang, "State-of-the-art review on FRP strengthened steel structures," Engineering Structures, vol. 29, no. 8, pp. 1808–1823, 2007.
16. H. Suzuki, "First application of carbon Fiber reinforced polymer strips to an existing steel bridge in Japan," in Advanced Materials for Construction of Bridges, Buildings, and Other Structures, 2005.
17. R. C. Allan, J. Bird, and J. D. Clarke, "Use of adhesives in repair of cracks in ship structures," Materials Science and Technology, vol. 4, no. 10, pp. 853–859, 1988.
18. Grabovac, R. A. Bartholomeusz, and A. A. Baker, "Composite reinforcement of a ship superstructure—project overview," Composites, vol. 24, no. 6, pp. 501–509, 1993.
19. A.Baker, "Fibre composite repair of cracked metallic aircraft components," in Proceedings of the Australian Aviation Symposium: 'Innovate or Enervate', Preprints of Papers, p. 107, and 1987.
20. A M. P. de Jesus, J. M. T. Pinto, and J. J. L. Morais, "Analysis of solid wood beams strengthened with CFRP laminates of distinct lengths," Construction and Building Materials, vol. 35, pp. 817–828, 2012.

21. X. Gu, B. Peng, G. Chen, X. Li, and Y. Ouyang, "Rapid strengthening of masonry structures cracked in earthquakes using fiber composite materials," Journal of Composites for Construction, vol. 16, no. 5, pp. 590–603, 2012.
22. M. Tavakkolizadeh and H. Saadatmanesh, "Repair of damaged steel-concrete composite girders using carbon fiber-reinforced polymer sheets," Journal of Composites for Construction, vol. 7, no. 4, pp. 311–322, 2003.
23. P. T. Curtis, "Fatigue behaviour of fibrous composite materials," Journal of Strain Analysis for Engineering Design, vol. 24, no. 4, pp. 235–244, 1989.
24. Y. Zheng, L. Ye, X. Lu, and Q. Yue, "Experimental study on fatigue behavior of tensile steel plates strengthened with CFRP plates," in Proceedings of the 3rd International Conference on FRP Composites in Civil Engineering (CICE ‹06), Miami, Fla, USA, December 2006.
25. Z. S. Wu, J. P. Ou, G. Wu, Y. L. Wang, and W. Q. Feng, "A type of continuous fiber-steel wires composite plate," China National Invention Patent, Publication no. CN101581133A, 2009, (Chinese).
26. Z. Wu, X. Wang, and G. Wu, "Advancement of basalt fiber composites towards infrastructural applications (Keynote paper)," in Proceedings of the International Symposium on Innovation & Sustainability of Structures (ISISS ‹11), October 2011.
27. S. Moy, "Early age curing under cyclic loading-an investigation into stiffness development in carbon fibre reinforced steel beams," in Proceedings of the 1st International Conference on Advanced Polymer Composites for Structural Applications in Construction, Southampton, UK, April 2002.
28. D. CNR, 202/2005. Guide for the Design and Construction of Externally Bonded FRP Systems for Strengthening Existing Structures, National Research Council, Rome, Italy, 2007.
29. O. Caglayan, K. Ozakgul, and O. Tezer, "Fatigue life evaluation of a through-girder steel railway bridge," Engineering Failure Analysis, vol. 16, no. 3, pp. 765–774, 2009.
30. Leander, A. Andersson, and R. Karoumi, "Monitoring and enhanced fatigue evaluation of a steel railway bridge," Engineering Structures, vol. 32, no. 3, pp. 854–863, 2010.

31. B O. Buyukozturk, O. Gunes, and E. Karaca, "Progress on understanding debonding problems in reinforced concrete and steel members strengthened using FRP composites," Construction and Building Materials, vol. 18, no. 1, pp. 9–19, 2004.

32. S. C. Jones and S. A. Civjan, "Application of fiber reinforced polymer overlays to extend steel fatigue life," Journal of Composites for Construction, vol. 7, no. 4, pp. 331–338, 2003.

33. M. Bocciarelli, P. Colombi, G. Fava, and C. Poggi, "Fatigue performance of tensile steel members strengthened with CFRP plates," Composite Structures, vol. 87, no. 4, pp. 334–343, 2009.

34. D Schnerch, M. Dawood, S. Rizkalla, E. Sumner, and K. Stanford, "Bond behavior of CFRP strengthened steel structures," Advances in Structural Engineering, vol. 9, no. 6, pp. 805–817, 2006.

35. B. EN, "Eurocode 3: Design of Steel Structures-Part 1-9: Fatigue," European Committee for Standardization (CEN), 2004.

36. B. N. B. C. BNBC, Housing and Building Research Institute, Bangladesh Standard and Testing Institute, 1993.

37. Y. J. Kim and K. A. Harries, "Fatigue behavior of damaged steel beams repaired with CFRP strips," Engineering Structures, vol. 33, no. 5, pp. 1491–1502, 2011.

38. G. Wu, H. Wang, Z. Wu, H. Liu, and Y. Ren, "Experimental study on the fatigue behavior of steel beams strengthened with different fiber-reinforced composite plates," Journal of Composites for Construction, vol. 16, no. 2, pp. 127–137, 2012.

39. U. Loher, B. Mueller, R. Leutwiler, and V. Esslinger, "CFRP-strengthened aluminum structures," in Proceedings of the 17th International SAMPE Europe Conference on Success of Materials by Combination, pp. 37–54, Basel, Switzerland, 1996.

40. H. Jiao, F. Mashiri, and X.-L. Zhao, "A comparative study on fatigue behaviour of steel beams retrofitted with welding, pultruded CFRP plates and wet layup CFRP sheets," Thin-Walled Structures, vol. 59, pp. 144–152, 2012.

41. S. Moy and F. Nikoukar, "Flexural behaviour of steel beams reinforced with carbon fibre reinforced polymer composite," in Proceedings of the Advanced Polymer Composites for Structural Applications in Construction: Proceedings of the 1st International

Conference, p. 195, Southampton University, Southampton, UK, April 2002.

42. Y. J. Kim and G. Brunell, "Interaction between CFRP-repair and initial damage of wide-flange steel beams subjected to three-point bending," Composite Structures, vol. 93, no. 8, pp. 1986–1996, 2011.
43. W. Sebastian and C. Zhang, "Observations from testing of indeterminate FRP-plated steel beams on flexible supports," Composites Part B: Engineering, vol. 45, no. 1, pp. 200–214, 2013.
44. M. A. A. Siddique and A. A. El Damatty, "Improvement of local buckling behaviour of steel beams through bonding GFRP plates," Composite Structures, vol. 96, pp. 44–56, 2013.
45. V. Rance and U. Evans, Corrosion and Its Prevention at Bimetallic Contacts, HM Stationary Office, London, UK, 1958.
46. B Brown, "The corrosion of CFRP-to-metal couples in saline environments," in Proceedings of the 2nd International Conference on Carbon Fibres, pp. 18–20, 1974.
47. M. Tavakkolizadeh and H. Saadatmanesh, "Galvanic corrosion of carbon and steel in aggressive environments," Journal of Composites for Construction, vol. 5, no. 3, pp. 200–210, 2001.
48. M. Dawood, S. Rizkalla, and E. Sumner, "Fatigue and overloading behavior of steel-concrete composite flexural members strengthened with high modulus CFRP materials," Journal of Composites for Construction, vol. 11, no. 6, pp. 659–669, 2007.
49. M. Dawood and S. Rizkalla, "Bond and splice behavior of CFRP laminates for strengthening steel beams," in Proceedings of the International Conference on Advanced Composites in Construction, University of Bath, Bath, UK, 2007.
50. A Yasin, Seismic performance assessment of reinforced concrete frame structures [M.S. thesis], BUET, Dhaka, Bangladesh, 2008.
51. L Hollaway and J. Cadei, "Progress in the technique of upgrading metallic structures with advanced polymer composites," Progress in Structural Engineering and Materials, vol. 4, pp. 131–148, and 2002.
52. D Schnerch, K. Stanford, E. Sumner, and S. Rizkalla, "Bond behavior of CFRP strengthened steel bridges and structures," in

Proceedings of the International Symposium on Bond Behaviour of FRP in Structures (BBFS ‹05), International Institute for FRP in Construction, 2005.

53. Monfared, K. Soudki, and S. Walbridge, "CFRP reinforcing to extend the fatigue lives of steel structures," in Proceedings of the Fourth International Conference on FRP Composites in Civil Engineering (CICE ‹08), Zurich, Switzerland, 2008.

54. J. Teng, D. Fernando, T. Yu, and X. Zhao, "Treatment of steel surfaces for effective adhesive bonding," in Advances in FRP Composites in Civil Engineering, pp. 865–868, 2011.

55. J. Deng and M. M. K. Lee, "Fatigue performance of metallic beam strengthened with a bonded CFRP plate," Composite Structures, vol. 78, no. 2, pp. 222–231, 2007.

56. J. Deng and M. M. K. Lee, "Behaviour under static loading of metallic beams reinforced with a bonded CFRP plate," Composite Structures, vol. 78, no. 2, pp. 232–242, 2007.

57. Schnerch, M. Dawood, S. Rizkalla, and E. Sumner, "Proposed design guidelines for strengthening of steel bridges with FRP materials," Construction and Building Materials, vol. 21, no. 5, pp. 1001–1010, 2007.

58. S. Choudhury, Effects of a debonding flew size and location on the flexural performance of FRP retrofitted steel sections [M.S. thesis], The University of Texas, Arlington, Tex, USA, 2007.

59. S. A. Hashim, "Adhesive bonding of thick steel adherends for marine structures," Marine Structures, vol. 12, no. 6, pp. 405–423, 1999.

60. A F. Harris and A. Beevers, "The effects of grit-blasting on surface properties for adhesion," International Journal of Adhesion and Adhesives, vol. 19, no. 6, pp. 445–452, 1999.

61. A A. El Damatty and M. Abushagur, "Testing and modeling of shear and peel behavior for bonded steel/FRP connections," Thin-Walled Structures, vol. 41, no. 11, pp. 987–1003, 2003.

62. N. Photiou, L. Hollaway, and M. Chryssanthopoulos, "An ultra-high modulus carbon/glass fibre composite system for structural upgrading of steel members," in Proceedings of the 2nd International Conference on FRP Composites in Civil Engineering (CICE ‹04), pp. 741–748, Adelaide, Australia, December 2004.

63. Ghafoori, A. Schumacher, and M. Motavalli, "Fatigue behavior of notched steel beams reinforced with bonded CFRP plates: determination of prestressing level for crack arrest," Engineering Structures, vol. 45, pp. 270–283, 2012.

64. E. Ghafoori, M. Motavalli, J. Botsis, A. Herwig, and M. Galli, "Fatigue strengthening of damaged metallic beams using prestressed unbonded and bonded CFRP plates," International Journal of Fatigue, vol. 44, pp. 303–315, 2012.

65. Z. Wu, X. Wang, K. Iwashita, T. Sasaki, and Y. Hamaguchi, "Tensile fatigue behaviour of FRP and hybrid FRP sheets," Composites B: Engineering, vol. 41, no. 5, pp. 396–402, 2010.

66. Wu, Z. Wu, Y. Luo, Z. Sun, and X. Hu, "Mechanical properties of steel-frp composite bar under uniaxial and cyclic tensile loads," Journal of Materials in Civil Engineering, vol. 22, no. 10, Article ID 010010QMT, pp. 1056–1066, 2010.

67. M. Phares, T. J. Wipf, F. W. Klaiber, A. Abu-Hawash, and Y.-S. Lee, "Strengthening of steel girder bridges using FRP," in Proceedings of the Mid-Continent Transportation Research Symposium, pp. 1–12, 2003.

68. F. Vatandoost, Fatigue behaviour of steel girders strengthened with prestressed CFRP strips [Master of Applied Science in Civil Engineering], The University of Waterloo, Waterloo, Canada, 2010.

69. P. Colombi, A. Bassetti, and A. Nussbaumer, "Delamination effects on cracked steel members reinforced by prestressed composite patch," Theoretical and Applied Fracture Mechanics, vol. 39, no. 1, pp. 61–71, 2003.

70. P. Colombi, A. Bassetti, and A. Nussbaumer, "Analysis of cracked steel members reinforced by pre-stress composite patch," Fatigue and Fracture of Engineering Materials and Structures, vol. 26, no. 1, pp. 59–66, 2003.

71. Soudki, A. Schumacher, and M. Motavalli, "Flexural strengthening of a steel beam with prestressed strips-preliminary investigation," in Proceedings of the 9th International Symposium on Fiber Reinforced Polymer Reinforcement for Concrete Structures (FRPRCS ‹09), Sydney, Australia, 2009.

72. Fernando, A. Schumacher, M. Motavalli, J. Teng, T. Yu, and E. Ghafoori, Fatigue Strengthening of Cracked Steel Beams with

CFRP Plates, American Society of Mechanical Engineering, Vancouver, Canada, 2010.

73. Täljsten, C. S. Hansen, and J. W. Schmidt, "Strengthening of old metallic structures in fatigue with prestressed and non-prestressed CFRP laminates," Construction and Building Materials, vol. 23, no. 4, pp. 1665–1677, 2009.
74. Zhou and T. L. Attard, "Rehabilitation and strength sustainability of fatigue damaged concrete-encased steel flexural members using a newly developed polymeric carbon-fiber composite," Composites Part B: Engineering, vol. 45, no. 1, pp. 1091–1103, 2013.
75. Zhou, T. L. Attard, B. Zhao, J. Yu, W. Lu, and L. Tong, "Experimental study of retrofitted reinforced concrete shear wall and concrete-encased steel girders using a new CarbonFlex composite for damage stabilization," Engineering Failure Analysis, vol. 35, pp. 219–233, 2013
76. H. Zhou, T. L. Attard, Y. Wang, J. Wang, and F. Ren, "Rehabilitation of notch damaged steel beams using a carbon fiber reinforced hybrid polymeric-matrix composite," Composite Structures, vol. 106, pp. 690–702, 2013.
77. James, "Optimising the fatigue resistance of materials," 1992.
78. Society, "Design guidance for strengthening concrete structures using fibre composite materials," Report of a Concrete Society Committee, 2000.
79. F. Nikouka, M. Lee, and S. Moy, "Strengthening of metallic structures using carbon fibre composites," in Proceedings of International Association for Bridge and Structural Engineering Symposium (IABSE ‹02), pp. 121–127, Melbourne, Australia, 2002.
80. P. E. Bourban, S. H. McKnight, S. B. Shulley, V. M. Karbhari, and J. W. Gillespie, "Durability of steel/composites bonds for rehabilitation of structural components," Proceedings of the 3rd Materials Engineering Conference Infrastructure: New Materials and Methods of Repair, pp. 295–302, 1994.
81. V. M. Karbhari and S. B. Shulley, "Use of composites for rehabilitation of steel structures—determination of bond durability," Journal of Materials in Civil Engineering, vol. 7, no. 4, pp. 239–245, 1995.

82. R. A. Barnes and G. C. Mays, "The effect of traffic vibration on adhesive curing during installation of bonded external reinforcement," Proceedings of the Institution of Civil Engineers: Structures and Buildings, vol. 146, no. 4, pp. 403–410, 2001.

83. S. Moy, "CFRP reinforcement of steel beams adhesive cure under cyclic load," in Proceedings of the 1st Asia-Pacific Conference on FRP in Structures (APFIS ‹07), pp. 1019–1024, 2007.

84. L. Zhang, L. Hollaway, J. Teng, and S. Zhang, "Strengthening of steel bridges under low frequency vibrations," in Proceedings of Composites in Civil Engineering (CICE ‹06), Miami, Fla, USA, 2006.

85. T. C. Miller, M. J. Chajes, D. R. Mertz, and J. N. Hastings, "Strengthening of a steel bridge girder using CFRP plates," Journal of Bridge Engineering, vol. 6, no. 6, pp. 514–522, 2001.

86. Narmashiri and M. Z. Jumaat, "Reinforced steel I-beams: a comparison between 2D and 3D simulation," Simulation Modelling Practice and Theory, vol. 19, no. 1, pp. 564–585, 2011.

87. Ghafoori and M. Motavalli, "Analytical calculation of stress intensity factor of cracked steel I-beams with experimental analysis and 3D digital image correlation measurements," Engineering Fracture Mechanics, vol. 78, no. 18, pp. 3226–3242, 2011.

88. Czaderski and O. Rabinovitch, "Structural behavior and interlayer displacements in CFRP plated steel beams—Optical measurements, analysis, and comparative verification," Composites B: Engineering, vol. 41, no. 4, pp. 276–286, 2010.

89. C. Czaderski, K. Soudki, and M. Motavalli, "Front and side view image correlation measurements on FRP to concrete pull-off bond tests," Journal of Composites for Construction, vol. 14, no. 4, pp. 451–463, 2010.

90. Y. Kawakam, H. Kanaji, and K. Oku, "Study on application of field signature method (FSM) to fatigue crack monitoring on steel bridges," Procedia Engineering, vol. 14, pp. 1059–1064, 2011.

91. Y. J. Xie, H. Xu, and P. N. Li, "Crack mouth widening energy-release rate and its application," Theoretical and Applied Fracture Mechanics, vol. 29, no. 3, pp. 195–203, 1998.

92. W. H. Müller, G. Herrmann, and H. GAO, "Elementary strength theory of cracked beams," Theoretical and Applied Fracture Mechanics, vol. 18, no. 2, pp. 163–177, 1993.
93. Y. J. Xie and X. H. Wang, "Application of G*-integral on cracked structural beams," Journal of Constructional Steel Research, vol. 60, no. 9, pp. 1271–1290, 2004.
94. G. Teng, T. Yu, and D. Fernando, "Strengthening of steel structures with fiber-reinforced polymer composites," Journal of Constructional Steel Research, vol. 78, pp. 131–143, 2012.
95. Wenguang and C. Guoping, "Analysis on fatigue life of cracked beam under forced vibration," in Proceedings of the 2nd International Conference on Computer Engineering and Technology (ICCET ‹10), pp. V546–V550, April 2010.
96. Z. Xiulin and M. A. Hirt, "Fatigue crack propagation in steels," Engineering Fracture Mechanics, vol. 18, no. 5, pp. 965–973, 1983.
97. R. Wang and A. Nussbaumer, "Modelling fatigue crack propagation of a cracked metallic member reinforced by composite patches," Engineering Fracture Mechanics, vol. 76, no. 9, pp. 1277–1287, 2009
98. P. Paris and F. Erdogan, "A critical analysis of crack propagation laws," Journal of Basic Engineering, vol. 85, p. 528, 1963.
99. S. Walbridge, "Fatigue analysis of post-weld fatigue improvement treatments using a strain-based fracture mechanics model," Engineering Fracture Mechanics, vol. 75, no. 18, pp. 5057–5071, 2008.
100. W. Johnson and S. Mall, "A fracture mechanics approach for designing adhesively bonded joints," ASTM STP, vol. 876, pp. 189–199, 1985.
101. A Kinloch and S. Osiyemi, "Predicting the fatigue life of adhesively-bonded joints," The Journal of Adhesion, vol. 43, no. 1-2, pp. 79–90, 1993.
102. S. Erpolat, I. A. Ashcroft, A. D. Crocombe, and M. M. Abdel-Wahab, "Fatigue crack growth acceleration due to intermittent overstressing in adhesively bonded CFRP joints," Composites A: Applied Science and Manufacturing, vol. 35, no. 10, pp. 1175–1183, 2004.

103. A. Ashcroft, "A simple model to predict crack growth in bonded joints and laminates under variable-amplitude fatigue," The Journal of Strain Analysis for Engineering Design, vol. 39, no. 6, pp. 707–716, 2004.

104. Lemaitre and R. Desmorat, Engineering Damage Mechanics: Ductile, Creep, Fatigue and Brittle Failures, Springer, 2005.

105. P. I. Kattan and G. Z. Voyiadjis, Damage Mechanics with Finite Elements: Practical Applications with Computer Tools, vol. 1, Springer, Berlin, Germany, 2002.

106. P. Raghavan and S. Ghosh, "A continuum damage mechanics model for unidirectional composites undergoing interfacial debonding," Mechanics of Materials, vol. 37, no. 9, pp. 955–979, 2005.

107. B. Bhattacharya and B. Ellingwood, "Continuum damage mechanics analysis of fatigue crack initiation," International Journal of Fatigue, vol. 20, no. 9, pp. 631–639, 1998.

108. M. A. Wahab, I. A. Ashcroft, A. D. Crocombe, and A. D. Crocombe, "Prediction of fatigue thresholds in adhesively bonded joints using damage mechanics and fracture mechanics," Journal of Adhesion Science and Technology, vol. 15, no. 7, pp. 763–781, 2001.

109. V. Shenoy, I. A. Ashcroft, G. W. Critchlow, and A. D. Crocombe, "Fracture mechanics and damage mechanics based fatigue lifetime prediction of adhesively bonded joints subjected to variable amplitude fatigue," Engineering Fracture Mechanics, vol. 77, no. 7, pp. 1073–1090, 2010.

110. L. Henry, A Theory of Fatigue Damage Accumulation in Steel, Ohio State University, 1953.

111. A. Bannantine, J. J. Comer, and J. L. Handrock, Fundamentals of Metal Fatigue Analysis, vol. 90, Prentice Hall, Englewood Cliffs, NJ, USA, 1990.

112. B. ANSI, AISC 360-05-Specification for Structural Steel Buildings, AISC, Chicago, Ill, USA, 2005.

113. Youssef, "Analytical prediction of the linear and nonlinear behaviour of steel beams rehabilitated using FRP sheets," Engineering structures, vol. 28, pp. 903–911, 2006.

114. H. Zhou, Y. Wang, Y. Shi, J. Xiong, and L. Yang, "Extremely low cycle fatigue prediction of steel beam-to-column connection by using a micro-mechanics based fracture model," International Journal of Fatigue, vol. 48, pp. 90–100, 2013.

115. A Pipinato, C. Pellegrino, and C. Modena, "Fatigue assessment of highway steel bridges in presence of seismic loading," Engineering Structures, vol. 33, no. 1, pp. 202–209, 2011.

116. S. Cheung and W. C. Li, "Probabilistic fatigue and fracture analyses of steel bridges," Structural Safety, vol. 25, no. 3, pp. 245–262, 2003.

117. P. Colombi, "Plasticity induced fatigue crack growth retardation model for steel elements reinforced by composite patch," Theoretical and Applied Fracture Mechanics, vol. 43, no. 1, pp. 63–76, 2005.

118. L. Chen, B. Weiss, and R. Stickler, "A model for crack closure," Engineering Fracture Mechanics, vol. 53, no. 4, pp. 493–509, 1996.

119. S. Stoychev and D. Kujawski, "Methods for crack opening load and crack tip shielding determination: a review," Fatigue and Fracture of Engineering Materials and Structures, vol. 26, no. 11, pp. 1053–1067, 2003.

120. J. C. Newman Jr, "A crack-closure model for predicting fatigue crack growth under aircraft spectrum loading," ASTM STP, vol. 748, pp. 53–84, 1981.

121. S. Wang and A. F. Blom, "A strip model for fatigue crack growth predictions under general load conditions," Engineering Fracture Mechanics, vol. 40, no. 3, pp. 507–533, 1991.

122. L. C. Hollaway and P. R. Head, Advanced Polymer Composites and Polymers in the Civil Infrastructure, Elsevier Science, 2001.

123. J. Cadei, T. Stratford, L. Hollaway, and W. Dcukett, Strengthening Metallic Structures Using Externally Bonded Fibre-Reinforced Polymers, vol. 595, CIRIA, London, UK, 2004.

124. Abed, Effects of Temperature on the Adhesive Bonding in Steel Beams Reinforced with CFRP Composites, University of Southampton, 2012.

125. T. Keller and M. Schollmayer, "Through-thickness performance of adhesive joints between FRP bridge decks and steel girders," Composite Structures, vol. 87, no. 3, pp. 232–241, 2009.

126. C.-T. Hoang-Ngoc and E. Paroissien, "Simulation of single-lap bonded and hybrid (bolted/bonded) joints with flexible adhesive," International Journal of Adhesion and Adhesives, vol. 30, no. 3, pp. 117–129, 2010.

127. M. Abdel Wahab, "Fatigue in adhesively bonded joints: a review," ISRN Materials Science, vol. 2012, Article ID 746308, 25 pages, 2012.

128. Kelly, "Quasi-static strength and fatigue life of hybrid (bonded/bolted) composite single-lap joints," Composite Structures, vol. 72, no. 1, pp. 119–129, 2006

129. S. Sam and M. Shome, "Static and fatigue performance of weld bonded dual phase steel sheets," Science and Technology of Welding and Joining, vol. 15, no. 3, pp. 242–247, 2010.

130. Moroni, A. Pirondi, and F. Kleiner, "Experimental analysis and comparison of the strength of simple and hybrid structural joints," International Journal of Adhesion and Adhesives, vol. 30, no. 5, pp. 367–379, 2010.

131. S. H. Yoon and D. G. Lee, "In situ crack propagation monitoring in tubular adhesive joints containing quartz nano-particles," Journal of Adhesion Science and Technology, vol. 25, no. 16, pp. 1973–1985, 2011.

132. Liu, X. Zhao, and R. Al-Mahaidi, "The effect of fatigue loading on bond strength of CFRP bonded steel plate joints," in Proceedings of the International Symposium on Bond Behaviour of FRP in Structures (BBFS ‹05), Hong Kong, China, 2005.

133. A Pipinato, C. Pellegrino, and C. Modena, "Fatigue behaviour of steel bridge joints strenghtened with FRP laminates," Modern Applied Science, vol. 6, no. 10, pp. 1–15, 2012.

134. K. T. Steudle and J. Horsley, AASHTO LRFD Bridge Design Specifications, 6th edition, 2012.

Citations

CHAPTER 1

Yeonho Park, Young Hoon Kim, and Swoo-Heon Lee, Long-Term Flexural Behaviors of GFRP Reinforced Concrete Beams Exposed to Accelerated Aging Exposure Conditions, doi: 10.3390/polym6061773.

CHAPTER 2

Wagner O. de Morais, Jens Lundström, and Nicholas Wickström, Active In-Database Processing to Support Ambient Assisted Living Systems, doi: 10.3390/s140814765.

CHAPTER 3

David J. LePoire, Review of Potential Characterization Techniques in Approaching Energy and Sustainability, doi: 10.3390/su6031489.

CHAPTER 4

Y.L. Mo and Rachel Howser Roberts (2013). Carbon Nanofiber Concrete for Damage Detection of Infrastructure, Advances in Nanofibers, Dr. Russel Maguire (Ed.), ISBN: 978-953-51-1209-9, InTech, DOI: 10.5772/57096.

CHAPTER 5

Remon Fayek Aziz and Sherif Mohamed Hafez, Applying Lean Thinking in Construction and Performance Improvement, doi:10.1016/j.aej.2013.04.008.

CHAPTER 6

Sreenivas Alampalli, and Arthur P. Yannotti, Designing Bridges for Inspectability, doi 10.1201/b17073-3.

CHAPTER 7

Sanjeev Kumar Verma, Sudhir Singh Bhadauria, and Saleem Akhtar, Probabilistic Evaluation of Service Life for Reinforced Concrete Structures, Chinese Journal of Engineering, vol. 2014, Article ID 648438, 8 pages, 2014. doi:10.1155/2014/648438.

CHAPTER 8

G.F. Sirca Jr, and H. Adeli, System Identification in Structural Engineering, doi:10.1016/j.scient.2012.09.002.

CHAPTER 9

Sergio Vincenzo Calcina, Luca Piroddi, and Gaetano Ranieri, Fast Dynamic Control of Damaged Historical Buildings: A New Useful Approach for Structural Health Monitoring after an Earthquake, http://dx.doi.org/10.1155/2013/527604.

CHAPTER 10

Mohamed Kamruzzaman, Mohd Zamin Jumaat, N. H. Ramli Sulong, and A. B. M. Saiful Islam, "A Review on Strengthening Steel Beams Using FRP under Fatigue," The Scientific World Journal, vol. 2014, Article ID 702537, 21 pages, 2014. doi:10.1155/2014/702537.

Index

A

B

C

D

E

F

G

H

R

S

T

U

V

W